Men at War

Politics, Technology and
Innovation in the
Twentieth Century

Men at War

Politics, Technology and Innovation in the Twentieth Century

Edited by

Timothy Travers
and
Christon Archer

Precedent
Chicago
1982

Library of Congress Cataloging in Publication Data
Main entry under title:

Men at war.

Includes bibliographical references.
Contents: Exerting control/Desmond Morton—Hindenburg, Ludendorff, and the crisis of German society, 1916-1918/Martin Kitchen—Mutiny in the mountains/Reginald Roy—[etc.]
1. Military history, Modern—20th century—Addresses, essays, lectures. 2. Military art and science—History—20th century—Addresses, essays, lectures. I. Travers, Timothy.
II. Archer, Christon I., 1940—
U42.M43   355'.00904   81-21178
                AACR2

Copyright © 1982 by Timothy Travers and Christon Archer.
All rights reserved.

Published by Precedent Publishing, Inc.
520 North Michigan Avenue, Chicago, IL 60611, USA.
ISBN 0-913750-21-2 (0-913750-46-8, paperback)
LC 81-80545

Contents

INTRODUCTION
    Tim Travers and Christon Archer   1

THE CIVILIAN FACTOR
    Exerting Control: The Development of Canadian Authority over the Canadian Expeditionary Force, 1914-1919   7
    Desmond Morton

    Hindenburg, Ludendorff and the Crisis of German Society, 1916-1918   21
    Martin Kitchen

    Mutiny in the Mountains: The Terrace, British Columbia Incident   49
    Reginald Roy

TECHNOLOGY AND TACTICS
    Sans Doctrine: British Army Tactics in the First World War   69
    Dominick Graham

    Aircraft versus Armour: Cambrai to Yom Kippur   93
    Brereton Greenhous

INTELLIGENCE
    Secret Operations versus Secret Intelligence in World War Two: The British Experience   119
    David Stafford

    Psychological Warfare and Newspaper Control in British-Occupied Germany: A Personal Account   137
    Frank Eyck

THE BOMBING WEAPON

    The Royal Air Force and the Origins of
    Strategic Bombing   149
        Sydney Wise

    The Development of Air Raid Precautions in Britain
    during the First World War   173
        Marian McKenna

PERCEPTIONS OF MILITARY HISTORY

    The Challenge of the Eighties: World War Two
    from a New Perspective, the Hong Kong Case   197
        Kenneth Taylor

    Learning Military Lessons from Vietnam: Notes
    for a Future Historian   213
        Joseph Ellis

Preface

Most of the essays in this volume originated from a lecture series titled War and Society that was held at the University of Calgary during 1977. Additional essays were invited to treat aspects of significant themes and to sum up the state and directions of research on modern military history in Canada. The editors gratefully acknowledge the assistance of contributors and the generous financial assistance of the following institutions and individuals: The Publications Committee, University of Calgary; Alberta Culture, Government of Alberta; The Royal Canadian Legion, Alberta; Mr. Arthur Child; and the Department of History, University of Calgary.

Contributors

Desmond Morton is Professor of History at Erindale College of the University of Toronto. Among his publications are *Ministers and Generals: Politics and the Canadian Militia, 1867-1904*; *The Last War Drum: The North-West Campaign of 1885*; and *The Canadian General: Sir William Otter*. He was President of the Canadian Historical Association, 1978-79.

Martin Kitchen is Professor of History at Simon Fraser University, and has written extensively on German history. His publications include *The German Officer Corps 1890-1914*; *A Military History of Germany from the Eighteenth Century to the Present Day*; and *The Silent Dictatorship: The Politics of the German High Command under Hindenburg and Ludendorff, 1916-1918*.

Reginald Roy is Professor of History and Strategic Studies at the University of Victoria. His most recent book is *For Most Conspicuous Bravery: A Biography of Major General G.R. Pearkes, V.C., Through Two World Wars*. He is currently working on the Canadian role in Normandy, 1944.

Dominick Graham is Professor of History and Associate Director of the Center for Conflict Studies, University of New Brunswick. He has published *Cassino*, and his *Masters of Fire Power: the British Army in Two World Wars*, with Shelford Bidwell, will be published in 1982.

Brereton Greenhous is a historian with the Canadian Department of National Defence. His interests center on tactical air power. He has written a number of articles on this topic and has lectured in Israel, West Germany, the United Kingdom, Canada and the U.S.A. He is also co-author of *Out of the Shadows: Canada in the Second World War*.

David Stafford is Associate Professor of History at the University of Victoria. He has worked for the British Foreign Office, has written on the European anarchist movement, and recently published *Britain and European Resistance 1940-1945*.

Frank Eyck is Professor of History at the University of Calgary. He served with the British Army in World War Two, subsequently worked for the British Broadcasting Corporation, and taught in Britain. His books include *The Frankfurt Parliament: 1848-1849*, and he has completed a biography G.P. Gooch.

Sydney Wise is Professor of History and Director of the Institute of Canadian Studies at Carleton University. He served in the R.C.A.F. in World War Two, and has written widely on Canadian and military history. His books include four editions of *Men in Arms*, and volume I of *The History of the Royal Canadian Air Force*. He was President of the Canadian Historical Association, 1974-1975.

Marian McKenna is Professor of History at the University of Calgary. Her written works range from *Borah: A Biography* to *Myra Hess: A Portrait*, and she is preparing for publication a manuscript on immigration to Canada.

Kenneth Taylor is Associate Professor of History in the Center for Strategic Studies and the Department of History, University of Alberta. After serving in the Royal Marines, he specialized in Soviet military studies at Princeton, and has written on a wide range of military topics. He is currently working on the Canadian forces in Hong Kong, 1941.

Joseph Ellis is Professor of History at Mount Holyoke College. He has taught at West Point and written extensively on American intellectual history, including *After the Revolution: Profiles of an American Culture*. He co-authored *School for Soldiers: West Point and the Profession of Arms*.

Christon Archer is Professor of History at the University of Calgary. His interests center on the Latin American military and he has published *The Army in Bourbon Mexico, 1760-1810*. He is working on counterinsurgency and on the Spanish exploration of the Northwest Coast of America.

Tim Travers is Associate Professor of History at the University of Calgary. He has written on British military thought and is preparing a manuscript on World War I.

For B.C.T., 1899-1979

Introduction

Timothy Travers and Christon Archer

The growing number of books on military history and the lively interest in military history courses at colleges and universities show that the study of war is enjoying considerable popularity in North America and elsewhere. The reasons for this are arguable, but of immediate interest is the kind of military history that is taught and written. Here the student of war comes across an interesting division of opinion as to how military history should be written. This dichotomy was aptly summarised some time ago by Robert Albion of Princeton: "Military history, lying as it does on the frontier between history and military science, requires a knowledge of both fields. This fact often presents a difficulty to the history teacher. . . ."[1]

In other words, military history was perceived as being the result of the combination of two separate disciplines. Sometimes this became too confusing to sort out, as when Alfred Vagts remarked that "war and battle are in many respects the outcome of neither art nor science. . .The record of battle is more or less confused at best."[2] The presumption of Vagts was that military history could be, and often was, written from either the civilian or military point of view, and that the reconciliation of the two was very difficult. Vagts' solution in 1937 was to civilianize war history "through a broader kind of military history." However, by 1958 Vagts was forced to conclude that the civilian histories of World War II resulted not in a "civilianization of war history, but rather in a militarization of civilian writers more anxious not to err on the side of military detail than to bring the latest of the wars into the larger frame of a general history of wars."[3]

Albion and Vagts had touched upon a particular problem that is not unique to military history. This problem relates to the fact that, generally speaking, history is a discipline by virtue of its subject matter, not by virtue of a particular methodology such as is characteristic of the sciences and of some social sciences.[4] There will be a variety of approaches to the writing of history, and Albion and Vagts were aware that military history was being written from at least two different points of view, namely a professional "internalist" approach and a civilian "contextual" view. This dichotomy is again not unique to military history, for the same dualism tends to occur in those areas of history, such as law and medicine, that can be written both by members of the profession concerned—lawyers and doctors—and by those outside the profession. Hence, one may contrast history written by lawyers, doctors and the military profession (the internalist view) with history written by "civilians" outside the profession (the contextual view). It is of interest to note that where a professional-civilian dichotomy does not exist, as in the history of ideas, for example, there tends to be a more generally unified approach to the subject.

The opposition of an internalist professional view and a contextual civilian approach can no longer be fully maintained, as Albion and Vagts once argued, since the boundaries between the two are no longer so clear and distinct. But their opposition suggests a useful "ideal" model with which to work, and for the purposes of this introduction internalist and contextual will continue to be contrasted in the sense that internalist carries the connotation of a more sharply defined emphasis upon weapons, tactics, and battle, and upon the professional interest of the military in technical aspects of warfare, while contextual carries the connotation of an emphasis upon the social, political and economic environment of the military. Neither of these two approaches is seen as having precedence over the other in terms of historical validity. It is maintained, however, that these two approaches represent fundamental traditions from which most current accounts ultimately stem in one way or another.

Most recently the emphasis in military history has been contextual, seeing military history "as a part of the whole of history, not isolated from the rest...the military as a projection of society at large, the relationships of the soldier and the state...military institutions and military thought."[5] Although this contextual approach now seems to be popular in military history, some recent discussions have re-emphasized the internalist interest in battle itself. In 1975 Russell Weigley noted that the new military historians, in stressing the contextual view of war, have failed to deal with the central fact of war, the battle itself: "Ultimately, conflict is what military history is all about..."[6] Similarly, in 1976 John Keegan argued that in the last resort military history is about the battle itself, rather than the wider context of the initiation, outcome and results of the battle. Thus Keegan wished to re-direct the contextual historians back to the central experience of conflict.[7] In fact both Weigley and Keegan were referring to what they considered an over-reaction of the new military historians against a traditional and largely discredited "drum and trumpet" school of battle and heroic war, which can be seen as an older internalist approach. The problem seems to be that at one extreme the contextual view can take the emotional content out of war, while at the other extreme the internalist view can put too much in. Weigley and Keegan would perhaps like to see a military history that combines the two approaches.

The discussion thus far has not been intended to show that there are only two ways of writing military history—the professional, internalist method and the civilian, contextual method—but that these are two fundamental sources from which a variety of ways of looking at war derive. A summary of the current situation in military history would therefore have to take account of the wide range of heirs and disciples of these two traditions. Stemming from the contextual side of the equation is the new military history—itself part of a wider resurgence of interest in the interactions of a total society, as advocated in the *Annales* school of French historians.[8] Somewhere between the contextual and internalist approaches lies the renewed interest in the experiential face of war, as exemplified by John Keegan and others of the "return to the battlefield" genre, such as Martin Middlebrook, *The First Day on the Somme*, 1971. Meanwhile, out in academic limbo still continues the "drum and trumpet" school, survivor of an older internalist tradition. A recent emphasis, although not a new topic, is the "logic of resources and technology" approach, such as Van Creveld, *Supplying War: Logistics from Wallenstein to Patton*, 1977, and the more popular Len Deighton, *Fighter: The True Story of the Battle of Britain*, 1977. Both of these can be seen as within the internalist area, although the real heirs to the professional "military science" of Albion and Vagts are probably the modern operations research and systems analysis fields.[9]

The content of *Men at War* reflects these strands. From the contextual viewpoint the section entitled "The Civilian Factor" demonstrates the interplay of politics, public opinion and military institutions. Professor Morton argues that Canadian control of the Canadian Expeditionary Force came about through the influence of politics, financial control, and military administrative competence. In a variant of the Fischer thesis, Professor Kitchen shows how the German High Command manoeuvred to retain its power and deflect revolutionary enthusiasm during the crisis of 1916-1918, while transferring the "lightning conductor" policy into the more active "stab in the back" theory. In the case of the 1944 Terrace, British Columbia mutiny, Professor Roy's research details the results of the awkward introduction of overseas conscription.

The "Technology and Tactics" section reflects something of the internalist approach in its discussion of weapons systems, tactics and ideas over the course of both World Wars. Professor Graham points out how weapons systems in World War I constrained tactical ideas and the ability of generals to plan, as well as showing that it was necessary for the British Army to fight, learn and re-train simultaneously. The paper by Mr. Greenhous analyzes the development of air power against armor over a lengthy period of time, and by the use of statistics makes a case for the effectiveness of air power over armor.

The "Intelligence" section returns to the contextual approach, and the two papers give a useful insight into the way that intelligence institutions operate. Professor Stafford reveals the dysfunctional rivalry of the Special Operations Executive and the Secret Intelligence Service and reinforces the contemporary impression that intelligence institutions are notoriously hard to control. Professor Eyck gives an eye witness account of his own work with psychological warfare and media control and implies that no other service operates so precisely at the difficult juncture of civil-military roles as does intelligence.

The "Bombing Weapon" section gives a contextual slant to the use of air power as a strategic weapon. Professor McKenna examines the success of Zeppelin raids on Britain during World War I, the civilian reaction, and the long process of preparation of air raid warning systems. Professor Wise weaves together the complex interrelations of events, theories, technology and civil-military discussions to show how the mystique of strategic bombing was produced, often to the detriment of other areas of military aviation.

The final section, "Perceptions of Military History," concerns the need for historians to step back and reevaluate, not just campaigns, but the ways in which military history is written and the ways in which wars are remembered, thought about, and analysed for lessons. In forthright manner, Professor Taylor looks at the official version of the defence of Hong Kong in late 1941 and finds it lacking. Outspoken in his criticism, Taylor calls for a new generation of military historians ready to look again at defeat as well as victory. Professor Ellis brings the volume up to date with a wide-ranging critique of the United States conception of war both before and during the Vietnam war, relating that conception to a structuralist view of the connection between war and society.

Ellis' paper also raises a final and very important military and historical question: do we really learn lessons from the past? Judging by historical experience the answer seems to be most often not, and so the military question becomes why not? Without engaging in the toils of the critical philosophy of history, one answer may be that military commanders and staff do not learn from the past because (to borrow Thomas Kuhn's suggestion)[10] each significant set of wars, and the social-economic structures that go along with them, produces a military "paradigm" of how war should be fought. The paradigm in force at any one time sets and solves military problems within its pre-determined boundaries, and so learning does not take place until a military revolution establishes a new paradigm. An example would be the Napoleonic paradigm of the 19th Century, only fully overturned during and after World War I by what might be termed the Technological paradigm of the 20th Century. While such a hypothesis leaves much to be worked out, this approach does something to combine the internalist and contextual traditions by placing the former within the latter, or by relating both traditions in an explicit manner.[11]

Notes

1. Robert Albion, *Introduction to Military History*, A.M.S. Press Reprint of 1929 edition, N.Y., 1971, p. vii. Maurice Matloff, Chief Historian of the U.S. Army Centre of Military History, makes the same point in almost exactly the same words: "Some Conclusions

about Military History," Russell Weigley, ed., *New Dimensions in Military History*, San Rafael, California, 1975, p. 388.
2. Alfred Vagts, *A History of Militarism*, revised edition, N.Y., 1967, p. 23.
3. *Ibid*, pp. 36-37.
4. Such is the argument of Jacques Barzun, "History: The Muse and Her Doctors," *American Historial Review*, vol. 77, Feb. 1972, pp. 55ff. Peter Paret also suggests the same when he calls for the use of psychology, economics, sociology, quantitative analysis, etc. by historians. By extension military history has need of the methods of other disciplines, presumably because it does not have a suitable methodology itself. Peter Paret, "The History of War," *Daedalus*, vol. 100, Spring 1971, pp. 376-396.
5. Matloff, *op cit*, p. 405; Russell Weigley, "Introduction," in Weigley, ed., *New Dimensions*, p. 11.
6. Weigley, *ibid*.
7. John Keegan, *The Face of Battle*, Penguin ed. 1978 (originally published London 1976), p. 28, and Chapter I generally.
8. It is doubtful whether the new military history appeared solely because of U.S. national security problems produced by W.W. II and the Cold War, as Russell Weigley claims: Weigley, "Introduction," *op cit*, p. 11.
9. If a methodology or set of methods assumes greater significance than the subject matter itself, then what is written ceases to be (military history) and becomes a separate field or discipline, just as various disciplines emerged from each other in the 19th Century. This may perhaps be the case for operations analysis in the 20th Century.
10. See the revision of Thomas Kuhn's *The Structure of Scientific Revolutions* (1962) by Lakatos and Musgrave, eds., *Criticism and the Growth of Knowledge*, Cambridge, 1970.
11. A useful analysis along these lines can be found in the work of Jacques Van Doorn, e.g. "The Decline of the Mass Army in the West: General Reflections," *Armed Forces and Society*, Winter 1975.

Exerting Control: The Development of Canadian Authority over the Canadian Expeditionary Force, 1914-1919.

Desmond Morton
Erindale College, University of Toronto

On October 14th, 1914, when men of the First Contingent of the Canadian Expeditionary Force reached Plymouth on the first stage of their journey to the Western Front, their status was clear. In the meaning of Britain's Army Act, the Canadians were "Imperial." They were soldiers of the British Army, recruited from the Empire. In the mood of the moment, any other status would have seemed inconceivable.[1] If there was any doubt of the full integration of the Canadians in the British Army, it was laid to rest by Canada's Minister of Militia, Colonel Sam Hughes: "we have nothing whatever to say as to the destination of the troops once they cross the water," Hughes told the Canadian House of Commons, "nor have we been informed as to what their destination may be."[2] In London, Canada's acting High Commissioner, George Perley, assumed, "that as soon as the Canadian troops arrive here they will be entirely under the authority of the War Office and become part of the Imperial army in every sense of the word."[3] In 1914 no one presumed otherwise.

Four years later, without formal or negotiated change of legal status, the presumption was very different. If the battle-hardened Canadian Corps remained under the operational control of the British General Headquarters, the Canadians now had their own commanders, organization and tactical doctrines. They were part of a Canadian, not an Imperial, army, under the effective authority of a chain of command which stretched through the Corps to a Canadian section at General Headquarters, a Canadian cabinet minister in London and a Canadian prime minister, Sir Robert Borden, who was insistent on a voice in the higher strategy of the war. The British commander-in-chief, Field Marshal Sir Douglas Haig, grumbled that the Canadians had come to see themselves as junior but sovereign allies. His political superior, the Earl of Derby, could only counsel resignation to altered circumstances: "we must look upon them in the light in which they wish to be looked upon rather than the light in which we would wish to do so."[4]

This transformation marked a crucial stage in Canada's march toward national sovereignty. Canada's military contribution to the Allied war effort, including the trauma of conscription and the sacrifice of 60,000 lives, was the foundation for subsequent marks of international standing, from the placing of signatures on the Covenant of the League of Nations to the famous Halibut Treaty of 1923. In the practical test of wartime administration, the fragile faith in imperial federation was also tested and demolished. The overseas experience of close to half a million Canadians drove home how much they held in common and how little of it they still shared with the people of the former mother country.[5]

Few historians have addressed the significance of Canada's developing autonomy during the years 1914 to 1918. Although Canada's independence from imperial ties became almost a blinding preoccupation in the postwar years, interpretation was dominated by disciples of a Liberalism convinced that nothing worthwhile could emerge from the Borden years. It has been left to Canada's military historians and to scholars of a later day to elucidate the events of the most decisive single decade in the growth of Canadian nationhood.[6]

Part of the mythology of Canada in 1914 was that the country was wholly unprepared for war. This illusion promoted Canadian self-esteem in the light of the rapid organization and despatch of the First Contingent. It also gave the Minister of Militia, Colonel Hughes, enormous and dangerous powers. In fact, Canada had an extensive mobilization plan which Hughes had scrapped on the eve of war. Her militia had expanded impressively in size and efficiency in the pre-war years. The country even had a precedent in the despatch of imperial expeditionary forces: in October of 1899, it took two weeks to recruit, organize, equip and despatch a thousand Canadians to the South African War.

Like the men of the Canadian Expeditionary Force, Lieutenant Colonel William Otter's battalion had become subject to British military law and command as soon as it reached Capetown. Unlike the soldiers of 1914, the volunteers of 1899 were almost an embarrassment to the government. Sir Wilfred Laurier would have been glad of a pretext to tell his Quebec supporters that Ottawa had done little more than transport some perfervid imperialists to an early fate. Instead, it was Lord Minto, the Governor General, who insisted that Canada's dignity demanded an organized regiment under Canadian officers.[7] It was Otter who insisted that Canadians would serve together as a unit, not as detachments scattered among the British regiments. It was Canadian public opinion that forced the Canadian government to establish a direct communications link between Ottawa and the commanders of successive Canadian contingents. The unfortunate Otter, squatting under a wagon in the pouring rain, dictating letters to his staff sergeant or composing reports with the stub of a pencil, was the forerunner of the monstrous overseas Canadian military administrations of two world wars.[8]

The South African experience taught lessons which would have to be learned again in 1914-18. Canadian soldiers in the field suffered from badly designed Canadian uniforms and equipment, complained about senior officers, British as well as Canadian, and discovered a surprising sense of national identity.[9] They returned from the war with a conviction of their military prowess, best expressed in the Militia Act of 1904. Henceforth, a Canadian officer could be good enough to command the nation's military force. Section 69 of the new act even envisaged the possibility that Canadians might serve beyond Canada's borders, although drafters of the section plainly envisaged no more than a hot pursuit of Yankee invaders or the seizure of the state of Maine.[10]

Throughout the British Empire, the South African War provoked a salutary increase in military professionalism and efficiency and a determined effort to standardize the training, tactics and staff doctrines of Britain and the dominions. Less successful in the face of Canadian economic nationalism were attempts to impose standardized weapons, uniforms and equipment. Canadian pride and entrepreneurial ambition produced the Ross rifle, the MacAdam shovel, the Oliver equipment, and boots that disintegrated in the mud of Salisbury Plain and Flanders.[11]

As it had remained throughout the nineteenth century, Canadian military planning continued to regard war with the United States as its primary official preoccupation. However, the prospect of sharing in imperial military adventures had always attracted militia enthusiasts. In the early 1890s, Major General Ivor Herbert had quietly reorganized and strengthened Canada's tiny permanent force so that it might some day be in a position to join its British counterparts.[12] The South African experience fostered the careful preparation of a mobilization plan for an overseas expeditionary force of an infantry division and a cavalry brigade. Designed by a gifted and articulate British staff officer, Colonel Willoughby Gwatkin, the scheme survived scrapping by Hughes to serve as at least an informal guideline in organizing both the first and second Canadian contingents. Gwatkin deserves more credit than he has ever received for mitigating the chaos created at Valcartier by the vain-glorious Colonel Hughes.[13]

In the light of his experience with his Canadian superiors, it is not surprising that Gwatkin made no prevision in his plan for continued or effective Canadian control of the expeditionary force once it left Canada.[14] Nor did Hughes. Apart from the famous but almost certainly mythical episode in which the Canadian minister allegedly browbeat Lord Kitchener into keeping the Canadians together, Hughes appeared at least initially content to allow his men to march into complete British control. It was South Africa all over again.[15]

However, there was one substantial but scarcely recognized difference between the arrangements of 1899 and those of 1914. In the earlier conflict, the Laurier government assumed no greater initial financial commitment than to raise, equip and deliver troops to the British base at Capetown. Thereafter the Canadians depended on the austere generosity of the British taxpayer. In 1914, in a mood of effulgent patriotism, the Borden government proposed, with unhesitating Liberal concurrence, that Canada bear the entire cost of the nation's military contribution. Parliament, the watchdog of the treasury, never barked. A Liberal back-bencher offered a vague query about the financial implications of such generosity but even he did not demand an answer. In due course, the commitment would add about a quarter of a billion dollars to the national debt.[16]

Far more than the National Debt was involved. Because Canada helped to pay for the music, the Borden government accumulated early and pressing reasons to criticize the orchestra. More important, the British government felt obliged to listen to the criticisms. The complaints began with the arrival in England in October, 1914 of a brash crony of Sam Hughes, Colonel John Wallace Carson. Armed with a vague mandate to watch over the comfort of the Canadians, Carson returned to Canada in December with reports of appalling conditions on Salisbury Plain and of the apparent refusal of the War Office and of the Canadian Expeditionary Force's British commander, Lieutenant General E.A. Alderson, to remedy the situation. Carson's reports awoke the first official Canadian reservations about British management of the common war effort.[17]

Carson soon returned to England, armed with the vague authority of the "Minister's representative," to denounce the wholesale replacement of Canadian boots, shovels, wagons, motor vehicles and other equipment by British equivalents. Both a financial and an economic issue were at stake. While the British soon made it clear that they would not present a bill for standardizing the equipment of the Canadian division, massive abandonment of Canadian material dashed hopes for rich war contracts. For a nation deep in economic recession, it seemed a savage and unnecessary blow.[18]

When most of the First Contingent left for France in February as the First Canadian Division, the remnants had to be organized as a base, receiving casualties, training and forwarding reinforcements, managing pay and records, and handling shipments of material and weapons, such as the Ross Rifle, which continued to come from Canada. From the first, the British withdrew from any attempt to oversee Canadian military affairs in England.[19] The War Office had preoccupations enough of its own. The outcome, in the absence of any single directing force, was an unedifying mess. By the autumn of 1915 no fewer than three Canadian senior officers could claim authority from the Minister of Militia to preside over all other Canadians. One of them was Carson. Advanced by his own ingenuity and relentless pressure to the rank of major general, armed with Hughes' personal friendship, and shameless in exaggerating his vague authority, Carson finally convinced the British that he was the proper channel for their communications.[20]

Any base of operations comes to be regarded with contempt by fighting soldiers but the Canadian operation in England was more contemptible than most. Some of the worst problems derived from Sir Sam Hughes' insistence on recruiting new battalions, commanded by politically influential but untrained officers, instead of finding more volunteers for existing units of the Canadian Expeditionary Force. Dozens of enthusiastic but untrained units reached England only to be dissolved. Subalterns and privates hurried off to France, leaving their seniors with a burning sense of grievance and a desperate urge to find a dignified post in the proliferating Canadian overseas organization. Such officers, like Carson himself, were ill-equipped to win the respect of fighting soldiers in France.[21]

Responsibility for the confusion belonged to the Canadian Minister of Militia. Hughes showed no inclination to resolve the conflicts of authority. When Sir Robert Borden authorized Hughes to deal directly with the British authorities, he was chagrined to discover that his Minister set the wires humming with abusive and often ill-informed messages. "You do not represent yourself alone as Minister of Militia," Borden warned, "but the Government of which you are a member."[22] To Kitchener, the Hughes eruptions were simply "extraordinary."

Sir Sam Hughes had a high opinion of his fellow amateurs, contempt for the professionals and no hestitation in making his views known: "It is the general opinion," he claimed to his friend, Max Aitken, "that scores of our officers can teach the British Officers for many moons to come."[23] However, that reputation was only forged painfully and at a high cost in lives at the front, where Hughes was firmly forbidden to meddle. In England, where his appointees squabbled and plotted, the Minister exploited the confusion to ensure that even trivial decisions were eventually referred to him in Ottawa. In no other way could he manage the affairs of his beloved Canadian Expeditionary Force in every detail. Compelled by the Prime Minister to find some better arrangement, Hughes' solution in the summer of 1916 was an "Acting Sub-Militia Council," staffed by Carson and a handful of dependable stooges, with the Minister's son-in-law to act as secretary and personal spy.[24]

Sir Robert Borden might, in Hughes' own view, be "as gentle-hearted as a girl,"[25] but the Nova Scotian had limits. In 1915 and 1916 large chunks of the Minister's authority had been assigned to new agencies and commissions. Throughout the summer of 1916 Borden waited impatiently for Hughes to return from England with a solution. Instead, the Minister had wiled away the weeks instructing recruits in bayonet fighting and visiting his beloved "boys." In September, furious at reports of the "Acting" council, Borden acted. For the first and last time, a government department would exist entirely outside Canada. Sir George Perley, Borden's dependable friend, would be Minister for the Overseas Military Forces of Canada. For two years he had been a critic; now he would be responsible for putting things right.[26] Infuriated at being supplanted by a life-long rival, Hughes resigned.

Perley later confessed that he would never have taken the job if he had understood all the problems.[27] He soon found that there was scarcely an aspect of Canadian administration in England, from the training camps to the chaplains' service, which did not fester in conflict, disorder, extravagance and outright corruption. Some of the quarrels, notably between the Canadian army medical service and its Hughes-appointed critic, Colonel Herbert A. Bruce, would reverberate long into the postwar world.[28] By dint of patience, common sense and the reputations of senior officers brought back from France to form his new staff, Perley achieved an impressive transformation. When Canadian soldiers overseas voted massively for the Borden government in 1917, the remarkable improvement in the overseas military administration was at least a factor.[29]

The work of Perley's new Overseas Ministry proved that Canadians could reverse a situation which, for two years, had stood in British eyes as proof of colonial unfitness for serious responsibilities. In Canada, Hughes might strike admirers as a two-fisted nationalist who spurned politics, red tape and military professionals. Overseas, the Minister had become a national embarrassment to all but the sycophants who depended on his favor.[30] In France, that favor could not withstand the test of battle. By December, 1916, with four Canadian divisions in the line, Perley could build his organization with officers who, for the most part, had won the confidence of the Canadian Corps.

In 1914 it had seemed inevitable that only a British officer would have the training and experience to command the Canadian division in the field. After admitting that the few Canadian-born generals in the British Army were too old, too incompetent or too far away, Hughes had accepted Sir Edwin Alderson, an active, conscientious professional who had successfully commanded Canadians in South Africa.[31] When a second division entered the line in September, 1915, Alderson was the natural choice to command the Canadian Corps. However, by vetting Alderson, the Canadian government had asserted its right to choose the officers it would pay and it could then insist all divisional commanders would thenceforth be Canadians.

The initial commander of the 2nd Canadian Division, Major General Sam Steele, was a venerable protege of the Hon. Robert Rogers and a sop to the pride of Canada's west. Obviously unfit for operations, Steele's removal was eased when the British offered him a command at Shorncliffe. This proved acceptable to Canadians only if a Canadian was chosen as Steele's successor.[32] Major General Richard Turner, selected despite Alderson's reservations, was a Quebec Conservative who had won a Victoria Cross in South Africa. In the 1st Division, Alderson's successor was Major General Arthur Currie, a Liberal from Victoria. A Canadian predilection for regional and political balance was satisfied.[33]

The principle of Canadian authority in the choice of senior commanders did not go untested. In the wake of the unsuccessful struggle for the St. Eloi craters, responsibility for the failure probably lay with Turner and one of his brigadiers. According to Currie, Turner had repeated an offence he had committed at the second battle of Ypres: failing to go forward to see for himself.[34] However, with Hughes' agent, Max Aitken, busy on the scene, it proved impossible to remove a Canadian officer, however culpable. Instead, it was General Alderson, already in the Minister's disfavor for criticizing the Ross rifle, who was removed. His successor was another British general, Sir Julian Byng.[35]

Byng's appointment, and perhaps some hidden misgivings about his own operational competence, may have helped persuade Turner to accept Perley's invitation to become chief of staff and chief military official in the new Overseas Ministry on December 1st, 1916.[36] When Byng was promoted to command an army after his brilliant success at Vimy Ridge, there was no question that his successor would be a Canadian; the only issue was which one. Turner was senior, but the British, for obvious and substantial reasons, preferred Currie. It was one of Sir George Perley's more substantial services that he recognized the merits of the stout ex-Liberal. It was also to Turner's credit that he accepted the frustration of his amibition with no more than a comparable promotion in rank and a formal declaration that he remained the senior Canadian officer overseas.[37]

There might well have been a Canadian-British conflict if Currie had not also been the Canadian choice. However, the complications in the appointment came not from the British but from fellow Canadians when Currie refused to allow his former friend, Garnet Hughes, to assume command of the 1st Division. Convinced that the ex-Minister's son was unfit for command in the field, Currie braved at least some political pressure from Borden and Perley.[38] The strain on the new Corps Commander was all the greater because, quite independently, he found himself threatened with exposure in a serious prewar financial scandal. Thanks to help from wealthy brother officers, remarkable discretion on the part of Perley and a tight circle of officials, and considerable luck, Currie weathered his crisis and went on to earn a high reputation as a field commander.[39]

With experienced and respected Canadian officers in command in the field and an efficient and increasingly coherent administrative base in England, the Canadian Corps was able to develop, by the summer of 1917, into a remarkable national institution. In a little-noticed passage in the order-in-council creating the Overseas Ministry, the appointment of a Minister had been justified by the claim that men in the Canadian Expeditionary Force were members of the Canadian Militia serving overseas, an explicit change from the Army Act relationship in 1914.[40] The change was unnoticed because nobody seemed directly affected. Canadianization was as gradual as the slow change of accents in batteries and battalions, when British-born "Old Originals" were killed or wounded and their places were filled by the Canadian-born. There was no sudden break with British methods or tactics. Until the end of the war, the key staff positions in the Corps and in each of the four divisions were filled by competent British staff officers. They contributed much of the painstaking planning and preparation which distinguished Vimy Ridge from so many disastrous counterparts on the Western Front. It was a style which became the trademark of the Canadian Corps.[41]

However, the Corps' reputation came increasingly to depend on the skill and ingenuity of citizen-soldiers like Brigadier-General A.G.L. McNaughton of the artillery, Brigadier General W.B. Lindsay of the engineers, and a dozen others who brought fresh experience to a form of warfare that often defied old precepts.[42] The Corps did develop some unique approaches: General Raymond Brutinel's motorized machine guns, for example, and General McNaughton's counter-battery techniques. What especially characterized the Corps, particularly under Currie, was a somewhat conservative determination to use the maximum weight of material in the hope of saving lives and winning objectives. The Canadian Corps used more artillery in longer barrages.[43] When, as Currie later alleged to Borden, British generals were using labor to build tennis courts, Canadian pioneers were adding still more barbed wire entanglements.[44] In the spring of 1918, when the British cut their infantry strength from twelve to nine battalions per division, Currie won a hard struggle (in which Garnet Hughes' ambitions were again a factor) to preserve his organization.[45] During the desperate March offensives, Currie could insist, successfully, that his Corps be held together when Sir Douglas Haig might well have used it piecemeal to hold back the German advance. The Corps, Currie insisted, had become unique.

> From the very nature and constitution of the organization it is impossible for the same liaison to exist in a British Corps as exists in the Canadian Corps. My Staff and myself cannot do as well with a

British Corps in this battle as we can with the Canadian Corps, nor can any other Corps Staff do as well with the Canadian Divisions as my own.[46]

If Currie could play such an independent role, it was in part because Canadian authority had crossed the channel. By establishing a firm structure for Canadian administration in England, the Overseas Ministry was in a position to extend its authority to Canadians in France. By securing command of the Corps for a Canadian, Perley had guaranteed that the commander's allegiance, unlike that of Byng or Alderson, would no longer be merely to General Headquarters. The problems was to link the Corps and the Ministry.

From the outset of the war, securing reliable accounts of the developments at the front had been the Borden government's most frustrating problem. After despatching a series of cronies, Sir Sam Hughes had turned to Max Aitken, the ex-Canadian millionaire and a man with unusual talents for intrigue and opinion management.[47] When Aitken turned from his Canadian Expeditionary Force services to the task of demolishing the Asquith government, he left behind in France a small liaison office at General Headquarters, headed by Colonel R.F. Manly Sims, and devoted largely to providing care and transportation for visiting Canadian dignitaries.[48]

Once again, Canada's financial contribution became a factor. In addition to the troops of the Canadian Corps and a detached cavalry brigade, the Dominion furnished a growing stream of men to work on military railways and in the forests first of Scotland and then of France. Since Canada paid the bills, this growing army of uniformed workmen was entitled to supervision by Canadians. In Britain, the Overseas Ministry could act; in France there appeared to be no suitable authority.[49]

With the formation of the Union Government, Sir George Perley had left the Overseas Ministry, preserving the illusion that the High Commissionership was an office of importance and well aware that he could never hold his neglected Quebec constituency in a general election. The new minister, transferred from the Militia Department after a year's experience, was a bustling Toronto manufacturer, Sir Edward Kemp.

Kemp found little to please him in his new department. London was cold, he could not get bacon for his breakfast and Perley had failed to leave him a proper office. In fact, Kemp concluded, Perley had deferred too much to his officers and they, in turn, had become dangerously oblivious of the political nature of their decisions. "The most of these men have been over here three years or more," Kemp complained to the Prime Minister, "and the more of military life they have seen, the less likely they are to appreciate such a thing as Public Opinion in Canada."⁵⁰ Kemp promptly insisted on an overseas equivalent to the Militia Council he had worked with in Ottawa. Next, he turned to improving his control over Canadians in France.

During the 1917 election Colonel Manly Sims had cheerfully compromised himself as the Union political agent in France. His successor would have to be more than a replacement. Even before Kemp's arrival, Perley had determined that all promotions, appointments, transfers and exchanges of officers would pass through the Canadian office at General Headquarters rather than through British channels. With this burden added to supervision of Canadian troops not with the Corps, a more influential officer would be needed. The outcome, after negotiations with a reluctant War Office and, at long intervals, with a suspicious but preoccupied Sir Douglas Haig, was the establishment of a Canadian Section at G.H.Q.⁵¹

Behind the innocent title was a significant principle. Through the head of the Canadian Section, Sir Arthur Currie could not only communicate directly with the Overseas Ministry but also, in special circumstances, with Haig himself. There could be no pretence that Currie commanded just another army corps. It was to the credit of both the British and the Canadians that they accepted the changed relationship with dignity and without disruption. It there was objection, it came from Currie, fearful lest the new section might become a barrier to his own channels of communication with Sir Edward Kemp. The solution came through a careful appointment. Brigadier General J.F.L. Embury had won Currie's confidence through brief service with the Corps; as a prominent Saskatchewan Conservative, he won the Minister's confidence; as a newly appointed judge, he could be depended on for dignity and discretion.⁵²

By the summer of 1918 the Canadian Corps had become, in many respects, an allied army, fully responsive to the administrative and political authority of its own government. The process was by no means complete. Unlike the Australians, Canadians accepted the rigors of British military law, including the harsh penalty of execution by firing squad.⁵³ Canadians made no effort to develop their own flying service until the final stages of the war. As part of the price of involvement in the Imperial War Cabinet, Borden felt obliged to respond to Lloyd George's demand for Canadian troops for such remote and questionable enterprises as the expeditions to Archangel, Murmansk, Baku and, most substantially, Vladivostok. Such was the price of playing a world role.⁵⁴

Such a role seemed appropriate to Borden because by 1918 Canada had become a nation with authentic military power. More than any other major formation in the British Expeditionary Force, the Canadians had been ready for the major breakthrough that occurred at Amiens on August 8th,

1918. Thanks to conscription, the Corps could play a spearhead role in the hundred days of advances which ensued. That, too, was part of the price.[55]

Is there a moral to this long and complicated story?

Perhaps there are two—one ancient, one modern.

The old and enduring moral is that self-government was secured for Canada as much by courage, competence and self-sacrifice as it was for any other great nation. Canada demonstrated its determination and capacity to manage its own affairs in fighting as an ally of the nation which had given Canada its traditions of government and freedom and one language.

The modern moral is that Canada paid her way. Canada entered the First World War as a self-governing colony but engaged in the war as a self-sufficient ally. Canada was neither a client state nor a dependent. Today, in a country whose armed forces lack weapons and equipment for their assigned roles, where numbers of trained men and women have dwindled steadily, and where it is assumed that a powerful and friendly neighbor will assume the major burden of Canadian security in a dangerous world, has Canada regressed a little from the maturity its ancestors had won for the country by November 11th, 1918?

Notes

1. A. F. Duguid, *Official History of the Canadian Forces in the Great War, 1914-1919,* General Series, vol. I, *Appendices*, pp. 24-5; Gwatkin to Christie, Oct. 1, 1914, Public Arcives of Canada (hereafter P.A.C.), Gwatkin Papers, M.G. 30 G. 13.
2. Canada, House of Commons, *Debates*, August 21, 1914, p. 56.
3. Perley to Borden, Sept. 18, 1914, P.A.C., Perley Papers, vol. I.
4. Derby to Haig, Nov. 2, 1917, in Robert Blake, *The Private Papers of Douglas Haig, 1914-1919* (London, 1952), p. 266; G.W.L. Nicholson, *Canadian Expeditionary Force, 1914-1919: The Official History of the Canadian Army in the First World War* (Ottawa, 1962), p. 381; R.A. Preston, *Canada and "Imperial Defense"* (Toronto, 1967), pp. 488-492.
5. See, or example, C.P. Stacey, "Nationality: The Canadian Experience," *Canadian Historical Annual Report*, 1968.
6. See for example, C.P. Stacey, "Nationality: The Canadian Experience," *Canada Historical World War I"* in J.L. Granatstein and R.D. Cuff, *War and Society in North America* (Toronto, 1971); Nicholson, *Canadian Expeditionary Force*; C.P. Stacey, *Canada and the Age of Conflict*, Vol. I, 1867-1961 (Toronto, 1977), pp. 172-239 passim.
7. D. Morton, *The Canadian General: Sir William Otter* (Toronto, 1974), pp. 161-5; Norman Penlington, *Canada and Imperialism, 1896-1899* (Toronto, 1965), chs. XVI, XVII.
8. "Lecture on the Paardeberg Campaign," p. 6, P.A.C., Otter Papers; Morton, *Canadian General*, pp. 206-8, 210; Toronto *Globe*, April 4, 1900.
9. Canada, Department of Militia and Defence, *Supplementary Report: Organization, Equipment, Despatch and Service of the Canadian Contingents During the War in South Africa, 1899-1900* (Ottawa, 1901), esp. pp. 11-13. See also Canada, House of Commons, *Debates*, Aug. 19, 1914, p. 16; W.H. McHarg, *From Quebec to Pretoria with the Royal Canadian Regiment* (Toronto, 1902).

10. Preston, *Imperial Defense*," p. 323ff; D. Morton, *Ministers and Generals: Politics and the Canadian Militia, 1868-1904* (Toronto, 1970), pp. 186ff.
11. Preston, *Imperial Defense*, pp. 40lff; Nicholson, *Canadian Expeditionary Force*, pp. 6-11. (All but the shovel might fairly be attributed to the Laurier government.)
12. Morton, *Ministers and Generals*, p. 102; Guy R. Machen, "The Canadian Offer of Troops for Hong Kong 1894," *Canadian Historical Review*, XXXVIII, 4, Dec. 1957.
13. Nicholson, *Canadian Expeditionary Force*, pp. 14-16; Duguid, *Official History: Appendices*, p. 11.
14. Duguid, *Offical History*, pp. 18-20; Gwatkin to Christie, Oct. 1, 1914, Gwatkin Papers.
15. Stacey, *Age of Conflict*, p. 178-9; Nicholson, *Canadian Expeditionary Force*, p. 35.
16. *Supplementary Report*, pp. 11-13, Canada, House of Commons, *Debates*, Aug. 21, 1914. See also Nicholson, *Canadian Expeditionary Force*, pp. 359-360. See Perley to Borden, Feb. 16, 1915, P.A.C., Borden Papers RLB 997 (1) pp. 117471 *et. seq.* for consideration of arrangements. P.A.C., Militia Department H.Q. 54-21-23-13, vol. 2; Nicholson *Canadian Expeditionary Force*, p. 361.
17. Borden to Perley, Jan. 12, 1915, Perley Papers, vol. 3. On Gwatkin's views, see Gwatkin to Christie, March 3, 1915, Gwatkin Papers; Perley to Borden, January 29, 1915, Perley Papers, vol. 3.
18. See Hughes to Borden, Sept. 1-2, 1916, Borden Papers, OC 308, pp. 33583-8.
19. Nicholson, *Canadian Expeditionary Force*, pp. 202-5.
20. See, for example, Carson to G.O.C., Southern Command, Feb. 18, 1915, cited in Nicholson, *Canadian Expeditionary Force*, p. 202; John Swettenham, *To Seize the Victory* (Toronto, 1965), p. 125-7.
21. Nicholson, *Canadian Expeditionary Force*, pp. 212-22; D. Morton, "French Canada and War, 1868-1917" in Granatstein and Cuff, *War and Society*, pp. 98-99; and D. Morton, "The Short, Unhappy Life of the 41st Battalion, C.E.F.," *Queen's Quarterly*, LXXXI, 1, Spring, 1974.
22. Borden to Hughes, June 8th, 1915, Borden Papers, OC 165 (2), p. 12696; cf. R.L. Borden, *Memoirs*, vol. II, (Toronto, 1930), pp. 462-3.
23. See Hughes to Aitken, November 30, 1915, Borden Papers, OC 318(1), p. 35582; Hughes to Kitchener, April 27, 1915, *ibid.*, OC 165 (2) p. 12693.
24. See D.M.A.R. Vince, "The Acting Overseas Sub-Militia Council and the Resignation of Sir Sam Hughes," *Canadian Historical Review*, XXXI, 1, March, 1950.
25. Hughes to McArthur, March 23, 1911, Borden Papers, OCA 36, v. 134.
26. Nicholson, *Canadian Expeditionary Force*, pp. 209-212.
27. Perley to Borden, Jan. 27, 1917, Borden Papers, OC 176, pp. 13638 ff.
28. See, for example, Sir Andrew Macphail, *The Official History of the Canadian Forces in the Great War, 1914-1918: Medical Services* (Ottawa, 1925) against Colonel Herbert A. Bruce, *Politics and the Canadian Army Medical Corps* (Toronto, 1919).
29. See D. Morton, "Polling the Soldier Vote: The Overseas Campaign in the 1917 General Election," *Journal of Canadian Studies*, X, 4, November, 1975; Perley to Borden, Dec. 10, 1917, Borden Papers, vol. 79, pp. 41143ff.
30. See, e.g., Leslie Frost, *Fighting Men* (Toronto, 1967), and Alan R. Capon, *His Faults Lie Gently* (Lindsay, 1970), defending Hughes.
31. Kitchener to Perley, Aug. 16, 1914, Perley Papers, vol. 1.
32. On Steele's position, see Steele to A.L. Sifton, Jan. 10, 1919, P.A.C., Sifton Papers, vol. 14; Perley to Borden, June 14, 1915, Perley Papers, vol. 4.
33. Alderson considered Turner unfit for command. See Alderson to Hutton, Aug. 21, 1915. British Museum, Add.MS:50096, Hutton Papers, p. 310. On Currie, see A.M.J. Hyatt, "The

Military Career of Sir Arthur Currie" (Duke University Ph.D. thesis, 1965); H.M. Urquhart, *Arthur Currie: The Biography of a Great Canadian* (Toronto, 1950).
34. Nicholson, *Canadian Expeditionary Force*, pp. 71-6; A.J.P. Taylor, *Beaverbrook* (London 1972), p. 89.
35. Nicholson, *Canadian Expeditionary Force*, pp. 146-7; Aitken to Hughes, April 26, 1916, Borden Papers, OC 183(2) pp. 14955ff.; Hyatt, "Currie," pp. 86-7.
36. Perley to Turner, November 24, 1916, Perley Papers, vol. 7; Turner to Perley, November 30, 1916, O.S. 11-78, p. 7186.
37. Nicholson, *Canadian Expeditionary Force*, pp. 283-4; Swettenham, *To Seize the Victory*, pp. 170-2; Gow to Perley, June 29, 1917, O.S. 10-8-7.
38. Hyatt, "Currie," pp. 83-6, 115-20; Swettenham, *To Seize the Victory*, pp. 170-3; Gow to Perley, June 29, 1917, O.S. 10-8-7.
39. Perley to Borden, July 21, 1917, Perley Papers, vol. 7; R.C. Brown and Desmond Morton, "The Strange Apotheosis of a Great Canadian: Sir Arthur Currie's Personal Crisis in 1917," *Canadian Historical Review*, Vol. LX, No. 1, March, 1979, pp. 41ff.
40. Duguid, *Official History, Appendices*, p. 5; Perley Papers, file 348 d./Nov. 1, 1916.
41. Swettenham, *To Seize the Victory*.
42. See, e.g., John Swettenham, *McNaughton*, vol. I, 1887-1939 (Toronto, 1968), pp. 49-165 *passim*; and, on Brutinel, Larry Worthington, *Amid the Guns Below: The Story of the Canadian Corps, 1914-1918* (Toronto, 1965); also Urquhart, *Carrie*, pp. 194-5.
43. On technique, see *inter alia*, Swettenham, *To Seize the Victory*, pp. 204-5.
44. Preston, "Imperial Defense," p. 490.
45. *Ibid.*, p. 202; Kemp to Borden, Feb. 8, 1918, P.A.C., Kemp Papers, vol. 182, file 62; Currie to Kemp, Feb. 7, 1918, Borden Papers, OC 494, pp. 52798-802.
46. Cited by Stacey, *Age of Conflict*, p. 195.
47. Duguid, *Official History; Appendices*, p. 161 [P.C. 29, January 6, 1915]; General Order 117, Sept. 23, 1915. On Carrick, see Carson to Hughes, July 9, 1915, P.A.C., R.G. 9, Carson File, 6-C-12.
48. Perley to War Office, Jan. 25, 1917, O.S. 10-8-1-.
49. Nicholson, *Canadian Expeditionary Force*, ch. XVI, pp. 485-90, 499-500.
50. Kemp-Borden, Feb. 24, 1918, Borden Papers, OC 485-D (1), p. 51153.
51. Minutes of a conference, April 2, 1918, O.S. 8-52. See also Borden papers, OC 485-D (2), pp. 51362-71, Lt. Gen. A.H. Lawrence to Overseas Ministry, June 23, 1918, O.S. 10-8-7; Embury memorandum, April 24, 1918, OS 10-8-7.
52. See O.S. 8-51, p. 5759; Currie to Kemp, July 13, 1918, O.S. 10-8-7. On Embury, Perley to Milner, June 15, 1917, *ibid.*
53. D. Morton, "The Supreme Penalty: Canadian Deaths by Firing Squad in the First World War," *Queen's Quarterly*, LXXIX, 3, 1972, pp. 348-357.
54. Stacey, *Age of Conflict*, pp. 276-284.
55. Preston, *"Imperial Defense,"* pp. 493-4.

THE CIVILIAN FACTOR

Hindenburg, Ludendorff and the Crisis of German Society, 1916-1918

Martin Kitchen
Simon Fraser University

One of the particularly fascinating aspects of the study of war and society is the examination of the ways in which the conflicts, divisions and contradictions within a particular society become intensified under the stresses of war to the point of social crisis.[1] In Germany the war was fought in part as a desperate attempt to overcome these social problems without making the necessary adjustments to the economic, political and social structures which were no longer adequate to the needs of a modern industrial society.[2] Failure to achieve a swift and decisive victory in the west and the onset of the war of attrition placed an almost intolerable strain upon a seriously divided class society that began to suffer the full horrors of the deprivation and agony of near total war.[3] Gradually it seemed to many observers that the war was about to provoke the social upheaval it was designed to avert, and calls for drastic remedial treatment became increasingly strident. It is within this context that the "Silent Dictatorship" of the High Command under Hindenburg and Ludendorff must be discussed.[4]

In his inaugural lecture at Freiburg in 1895 Max Weber warned of the dangers to a society when an economically declining class held political power, and spoke of an "economic death struggle" of the Prussian Junkers.[5] This theme was taken up twenty years later by Thorstein Veblen who analysed the curious combination of a modern and rapidly developing industrial sector and the continued domination of society by pre-industrial elites who had preserved their position in spite of the constitutional changes of 1867 and 1871.[6] Both Weber and Veblen were developing a train of thought started by Karl Marx who had written of Germany in 1867 "le mort saisit le vif."[7]

The unification of Germany by "blood and iron" occurred at a time when the industrialisation of the country was beginning to have a profound impact on the structure of society, even though that industrialisation was highly uneven and regionalised.[8] The constitution of the new Reich made possible the domination of Germany by Prussia and of Prussia by its landowning aristocracy, in spite of the concessions which appeared to have been made to parliamentary democracy. Although many conservatives feared that Bismarck had made far too many concessions to liberalism by creating a Reichstag elected by universal suffrage, the chancellor argued that he had deliberately used "parliamentarism to destroy parliamentarism" and was brilliantly able to manipulate the constituent parts of the new Reich to establish his particular brand of Bonapartist dictatorship.[9] Universal suffrage helped to placate the liberal bourgeoisie who were enjoying the benefits of the extraordinary economic boom of the *Grunderjahre* from 1871 to 1873, but it soon became apparent that the Reichstag was but the "fig-leaf of absolutism" and that the new suffrage had brought few of the benefits that its advocates believed were its necessary consequence.

With the onset of the "Great Depression" in 1873 the Bismarckian system began to assume its final form.[10] The agrarians began to demand protective tariffs and the way was open for the alliance of "rye and iron", of "chimney-stack barons and cabbage-junkers", of the landowning aristocracy and the industrial bourgeoisie. This marriage of convenience could only be maintained by constant harping on the imminent danger of socialism. Anti-socialism was thus a vital constituent part of Bismarck's system, not merely to counter a perceived threat to the established order, one likely to come not by revolution but by a progressive democratisation of the political life of the country. Anti-socialism also was a means of making a clear distinction between insiders and outsiders and thus of strengthening the ties within the in-group by constant vilification of its real and imagined enemies. This system of negative integration was maintained by hysterical denunciations of the "enemies of the Reich", of the social-revolutionary cancer of the socialists, of the republicanism of the left-liberals, of the ultramontanism of the catholics, the internationalism of the Jews and the anti-German attitudes of Poles and Alsatians.[11] In the short term it may have helped to maintain the conservative alliance and to win support from the industrial and commercial bourgeoisie for a system that was still dominated by the old elites, but it also served to intensify the tensions and contradictions it was designed to contain and control. In the long run the continued insistence that the world was divided into two simple categories of friend and foe was to lead directly to the evil practices of German fascism. Barricaded against the "enemies of the Reich", the "cartel of the state-preserving and productive estates" were to defend the Bismarckian system against unwanted change and to combat at every turn the emancipatory tendencies and movements of an industrial society.[12]

The social and political status quo was painfully maintained by the manipulative use of foreign policy and by imperialist expansion. The worst effects of the depression and of the high food prices caused by protection could be partially counteracted by social policy measures. The problems of an advanced capitalist economy suffering the effects of cyclical depression could be rendered less severe by an increasing level of state intervention in economic affairs. But the fundamental issues of the day were avoided.[13] The "red peril" became less of a manipulative slogan and more of a frightening possibility. The social democrats' revolutionary Marxism had always been of dubious sincerity, and by the end of the century it had been further watered down by the revisionist practice of the party. But social democracy was undoubtedly the major modernising and emancipatory movement in Imperial Germany. It continually gathered mass support, and was gaining the approval of liberal and democratic elements. The spectacular success of the party in the Reichstag elections of 1912 prompted further frantic efforts to patch up a powerful alliance on the right to resist any further advances by the progressive forces.[14] Thus Bismarck and his successors had two alternative policies, either repression by means of anti-socialist legislation or in the final resort a coup d'etat that would destroy the constitutional structure that seemed to be failing in its Bonapartist function, or to concentrate the bourgeois parties in the defence of the existing order—the policy of *Sammlungspolitik*.

A coup d'etat was too risky, but social imperialism and *Sammlungspolitik* were miserable failures. The recovery of the world economy in 1896 did not solve Germany's difficulties. Germany had become a great industrial power but it was desperately short of capital, and chronically so after the Agadir crisis when French capital was no longer available in Berlin. Overseas trade was hampered by protectionism, and its dependence on foreign sources of raw materials was increasingly a source of concern. Admiral Tirpitz's massive naval building programme had given many valuable contracts to the industrialists, but it had also further poisoned international relations. The fleet and the army were placing an almost intolerable strain on the financial resources of the Reich, and the determined opposition of the Junkers was so strong that a long overdue tax reform could not be implemented.[15] The vast cost of the military establishment was paid in large part by a highly regressive system of indirect taxation, and the landowning aristocracy continued to enjoy extraordinary privileges which prompted even the Chancellor, Bulow, whom none could accuse of liberal tendencies, to denounce them as "lousy egoists". Without the resources to increase military expenditure the army and the navy would lose their integrating function in domestic politics and might well be unable to match the improvements in the armies and navies of Germany's potential enemies.[16] Without a successful imperialist policy there could be little hope of the "nationalisation of the masses".[17] German policy by 1914 seemed to be charging "full steam ahead" down a dead-end street. To many in government circles the assassination at Sarajevo created a situation in which Germany was given a golden opportunity to solve its most pressing economic, social and political problems in a desperate *va banque* play. It was now or never.

At first it seemed that the war was providing the solutions to the problems of the last forty years. The Kaiser had announced that he no longer recognised parties and only knew Germans, and the political truce of the *Burgfrieden* seemed to hold up at first with the news of Germany's military victories and then, when the military situation became deadlocked, with the attractive prospect of massive war aims which would make the suffering of the moment well worthwhile. The country seemed to be united in the "community of field-grey" and the divisions and struggles of the past appeared to be forgotten in the common struggle. Yet the overriding problem was that if the war was to do the trick it had not only to be successful but also relatively swift. As a social imperialist tactic a long and protracted war was unlikely to be successful, and in a lengthy war of attrition, Germany, which lacked adequate sources of raw materials and had made inadequate financial preparations for war, was no match for the Entente with its much greater supplies of men and materials. Falkenhayn, von Moltke's successor as chief of the general staff after the failure of the Schlieffen plan at the battle of the Marne, was unable to punch a hole through the Entente's line at the battle of Ypres, and his attack on Verdun, although highly ingenious, ended in a bloody stalemate. In the summer of 1916 the British army was pressing hard against the German positions on the Somme. The Austrians suffered serious defeats against the Russians, who occupied Southern Galicia and the Bukovina after the early successes of the Brusilov offensive. The Italians were also successful against the Austrians, taking Gorizia in August.

The appointment of Hindenburg as chief of the general staff on 28 August 1916, with Ludendorff as his righthand man, was the result of a complex intrigue in which many levels of government, the military and the politicians were involved.[18] News of the change at the High Command (O.H.L.) was greeted with almost unanimous enthusiasm. Hindenburg and Ludendorff were the glorious victors of Tannenberg whereas Falkenhayn was responsible for the failure of the Verdun offensive. A busy propaganda machine had built up Hindenburg into a figure of almost godlike stature; most Germans were ready to believe that he would bring decisive victory and restore the unity of the nation that was falling apart after two years of war.[19] A "Hindenburg peace" would achieve the war aims that would satisfy the rapacity of the annexationists and provide the rich rewards that were deemed necessary for the people to become reconciled once again to the status quo. For the first time in the history of the Germany army the appointment of a chief of the general staff had been the topic of lengthy and excited public debate; and yet the Kaiser was almost alone in realising this pseudo-democratic element in the affair. The army had pushed their "Supreme Warlord" aside in August 1914, and now it seemed that the people were demanding a voice in army appointments, which were constitutionally the exclusive right of the monarch and one of the most important of royal prerogatives. William II was quite wrong to imagine that the appointment of Hindenburg and Ludendorff had anything to do with democracy, but he was perhaps justified in his concern that the manipulative and quasi-plebiscitary aspects of the Bonapartist system of domination were becoming too obvious. Bismarck had been careful to disguise his semi-absolutism in an exaggeratedly hypocritical loyalty to the crown. He was to insist that he was the humble vassal of the Elector of Brandenburg, and although William I sighed that it was indeed difficult to be Kaiser with such a chancellor, he knew that the monarchy had a significant role to play in the Reich. Hindenburg and Ludendorff lacked this subtle understanding of the manipulative possibilities of a monarchist system. They had little respect for William II and were prepared to use the fact of their exceptional popularity to claim to represent the will of the people in a way that the Kaiser "by the Grace of God" would never even consider.

The problems facing the new High Command were awesome. They were not only called upon to develop a new strategic approach to the war that would end the deadlock in the west and bring final victory, they were also faced with severe economic problems of war production, rapid inflation and mounting hardship and discontent. Lastly, with the growing disenchantment with the Chancellor, Bethmann-Hollweg, they were to be thrust into the centre of the political arena and were to devote as much of their time to political intrigue as to the task of conducting the war.

While at High Command East, Hindenburg and Ludendorff had imagined that the answer to the overall strategic question was simple. Falkenhayn had failed to have his Tannenberg and was therefore bitterly jealous of his rivals in the east. If only reserves could be sent from the western front to the east, a second massive battle of encirclement could be fought that would force Russia out of the war. Then the full might of the German army could be thrown against the western front, where the war would be won. The Austrian chief of staff, Conrad, supported this view and hoped to be able to mount a full scale offensive in Italy if only Falkenhayn would spare the troops. Then the Austrian army could join in the triumphant victory march in the west. Hindenburg and Ludendorff's belief in the decisive battle was attractively simple and in the traditional line of general staff thinking. The decisive battle would end the war quickly, achieve the war aims and solve the social and political problems that were becoming increasingly acute. Falkenhayn, however, continued to argue that the idea of a decisive battle was unrealistic. Such a battle could not be fought on the western front with the resources available to the Central Powers. Reserves could not be spared for the east because German troops on the western front were stretched to the limits of their endurance, and any further weakening of the front would lead to a serious risk of an enemy breakthrough. The only hope was that the Entente could be gradually weakened by the superior military skills and careful management of men and materials by the German troops. In the final analysis both positions were unrealistic. Germany lacked the resources for a successful war of attrition, even though the Entente powers were suffering terrible losses in their offensives; the strategy of the decisive battle ignored the very serious situation on the western front.

On their arrival at headquarters Hindenburg and Ludendorff were soon to find that Falkenhayn had been quite correct and that it was not for reasons of selfish jealousy that he had denied their repeated requests for reserves. Without the reserves there could be no talk of a decisive battle in the east. Efforts to recruit a Polish army to make up the necessary strength were a spectacular failure. The proclamation of an independent Poland was made on 5 November 1916, Hindenburg and Ludendorff having dismissed all the arguments that an independent Poland under German control would ruin the chances of a negotiated peace with Russia, would poison relations with Austria-Hungary, and might establish a dangerous precedent for other nationalities. Julian Marchlewski wrote: "Today's improvisation on the Vistula by Hindenburg and Ludendorff is unique, a joke the like of which the world has neither seen nor dreamed. An 'independent' state with unknown frontiers, with unknown government, with an unknown constitution and oh horror, oh shame, a kingdom without a king!"[20] In these circumstances it is hardly surprising that the call for volunteers for a Polish army under German control met with very little response. Slogans such as "No Polish army without a Polish government!" were used to discredit the recruiting drive. Less than five thousand applied, many of whom were found to be unsuitable for military service. The German governor of occupied Poland, Beseler, had argued in October that the Poles would willingly supply thirty-six battalions of volunteers within eight months of the declaration of independence.[21] Within a few weeks it was plain for all to see that these hopes were based on little more than wishful thinking.

With the realisation of the critical situation on the western front and with the failure of their Polish policy, Hindenburg and Ludendorff had to find some way out of the military impasse. For some time a very powerful group, with representatives from the conservative party, the national liberals, and the centre party, acting as spokesmen for the interests of the agrarians, heavy industrialists and Pan Germans, had argued in favour of unrestricted submarine warfare. Hindenburg and Ludendorff met with the navy leaders shortly after their appointment to the OHL to discuss the issue. Much to the disappointment of Holtzendorff, the chief of the admiralty staff, and Capelle, the state secretary of the navy, the army leaders argued that unrestricted submarine warfare should not begin before Romania had been conquered, for the risk of an attack from neutral Denmark and Holland, which posed a real threat to Germany's flank, could not be taken until reserves could be withdrawn from Romania to defend the borders in the north. Hindenburg and Ludendorff's motives were largely selfish. They wanted a spectacular success in Romania to further strengthen their position, and did not want the navy to steal their thunder in their first weeks at the High Command. It is difficult to imagine that they were seriously worried about the threat from Denmark and Holland, but this was a useful excuse and one which Bethmann-Hollweg seized upon in his opposition to the immediate opening of unrestricted submarine warfare. On the other hand Hindenburg and Ludendorff agreed at this conference at Pless that unrestricted submarine warfare was desirable at some later date; they had not rejected the idea itself.[22] The navy kept up its pressure on the OHL in the following weeks, and when German troops entered Bucharest on 6 December 1916 Hindenburg and Ludendorff could no longer object to unrestricted submarine warfare on military grounds.

On 8 December 1916 Hindenburg informed the Chancellor that unrestricted submarine warfare should begin at the end of January.[23] Having made this decision the High Command pushed consistently for unrestricted submarine warfare, ignoring Bethmann's attempts to use a peace move as diplomatic cover for the escalation of the war at sea and denouncing all attempts to negotiate or to seek a compromise as likely to lead to a "feeble peace" when only a "Hindenburg victory" could give Germany its rightful place in the world. The navy argued that within five months of the beginning of unrestricted submarine warfare 39% of shipping en route for Britain would be sunk, two-fifths of the neutral ships going to Britain would no longer be prepared to risk the voyage, and 600,000 tons of shipping would be sunk each month.[24]

The decision in favour of unrestricted submarine warfare was made at Pless on 9 January.[25] Bethmann had no defence against the military arguments: the dubious statistics of the navy, and Hindenburg's insistence that unrestricted submarine warfare was vital to restore the morale of the troops and to relieve the pressure on the Somme. The Entente would then be unable to mount a fresh offensive in the spring and England would be brought to its knees within a few months. Even Admiral Muller, a man noted for his caution, imagined that the war would be over by August.[26] In the mood of euphoria over the decision on submarine warfare there were few who stopped to think of the terrible risk they were running if America entered the war. Although Admiral Capelle told the budgetary committee of the Reichstag that this did not matter because every American troopship would be sunk, this too was an idle boast. Not one troopship was lost, and the navy's statistics were soon shown to be hopelessly unrealistic. Hindenburg and Ludendorff's submarine policy was as great a failure as Falkenhayn's strategy at Verdun, and they were unable to find any way of achieving the decisive victory which they had claimed only a few weeks previously had been denied to them by Falkenhayn's perverse behaviour.

The February revolution in Russia offered the possibility of some change in the situation in the east. At first it was hoped that the revolution would cause Russia to leave the war immediately, but the Lvov-Miliukov government determined to continue the war in spite of the obvious war-weariness of the Russian people. The OHL (High Command) supported the idea of sending Lenin back to Russia to stir up revolution so that the Russian war effort would be undermined.[27] Lenin travelled through Germany on the night of 10 April, by which time the Germans had virtually ceased their military operations in the east in an attempt to convince the Russian people that their desire for peace was genuine. General Hoffmann, the chief of staff in the east, compared the use of propaganda tactics in warfare to the use of grenades and poison gas, but Lenin and the Bolsheviks were dangerous weapons for the Germans to use. The armistice was agreed upon by 15 December and the negotiations for a peace settlement began at Brest Litovsk on 22 December, but the impact of the October revolution on the German left was considerable. The Bolshevik victory lead to a considerable radicalisation of the German working class and a further heightening of domestic political tensions.[28] Furthermore, the OHL used the weakness of the new regime in Russia to pursue its far-reaching aims in the east, and as they also were determined to oust the Bolsheviks from power once they had outlived their usefulness, large numbers of German and allied troops were still tied down in the east. Thus peace in the east did not lead to the decisive strengthening of the forces in the west. Germany still lacked the resources to fight a decisive battle on the western front.

The failure of Hindenburg and Ludendorff to find a military solution to the problems that had eluded Falkenhayn forced them to admit as early as November 1916 that another Tannenberg was out of the question and that the war was indeed one of attrition, a fact that they had vehemently denied only two months before. Even before their appointment to the OHL, Hindenburg and Ludendorff had been in close contact with those leaders of heavy industry who favoured a closely controlled war economy. At the OHL under Falkenhayn, the head of section 11, Colonel Bauer, who was one of the most outspoken supporters of the appointment of Hindenburg and Ludendorff, had also established close ties with leading industrialists and was an enthusiastic protagonist of a controlled economy. On their arrival at Headquarters Hindenburg and Ludendorff were immediately approached by leading industrialists, including Krupp, Duisberg and Rathenau, all of whom called for determined measures to increase the production of war materials.[29]

The new OHL had little interest in the debate that had been going on about a possible "economic general staff", which some industrialists saw as a threatening move towards "state socialism", others as a necessity, and a third group as an unfortunate but unavoidable consequence of the war.[30] To Hindenburg and Ludendorff it was simply a matter of maximizing war production to overcome the material superiority of the Entente. Hindenburg called for a threefold increase in the production of machine guns and artillery by the following spring along with a one hundred per cent increase in the production of ammunition.[31] If necessary, non-essential industrial plant would have to be closed down so that these targets could be met. The OHL felt that this declaration of intent was not enough and that the "Hindenburg Programme" should be supported by an "Auxiliary Labor Law" (*Hilfsdienstgesetz*) to be passed by the Reichstag in what was to be an act of national dedication, a total commitment by the German people to victory, and a dramatic demonstration to the Entente powers of Germany's determination to win the war.

Hindenburg informed the Chancellor of his ideas about an auxiliary labor law in a memorandum of 13 September 1916.[32] He called for an increase in the age of those liable for military service to fifty and a drastic reduction in exemptions. Industrial workers were to be moved away from non-essential industries into war production, where they should be subjected to "maximum exploitation". The good old principle of "he who does not work shall not eat" should be reintroduced, and childless women "who only cost the state money", and children who had nothing useful to do, should be forced to work for war industry. All pessimists, grumblers and profiteers were to be summarily and severely punished. A labor office (*Arbeitsamt*) should also be formed which would control wages, direct labor and withhold food from those who had no acceptable excuse for not working.

These drastic suggestions for the militarisation of the economy were designed to end any attempt at cooperation between capital and labor, to destroy the last vestiges of competitive capitalism and to undertake a major structural change in the German economy in favor of heavy industry. Disgust at war profiteers and blackmarket operators, alarm at a steadily deteriorating economic situation, and the vivid contrast between the sacrifices of the frontline soldiers and the easy life of many civilians had made the corporatist ideas of a "communal economy" (*Gemeinwirtschaft*), which was contrasted to "egotistical capitalism" and the "unbridled private economy", common currency by 1916.[33] Apostles of "*Gemeinwirtschaft*" trotted out Frederick the Great, Fichte, Stein, List, Bismarck and Lagarde as their intellectual ancestors. Troops in the trenches were provided with suggestive quotations from these mighty figures to convince them that they were fighting for a new, just and uniquely German economic system.

Bethmann-Hollweg was far from enthusiastic about these proposals. It was not so much that the target figures were hopelessly unrealistic and that there was no point in forcing women and children to work when there were no available places for them in the factories, but rather that he had no desire to hand over control of "all matters of war work, food and the production of war materials" to the military and thus further diminish the power of the civil authorities. His economic advisor, Helfferich, opposed the idea of a rigorously controlled economy, saying, "One can command an army, but not an economy."[34] Both the Chancellor and Helfferich believed in a degree of cooperation between capital and labor. They feared that such drastic measures would make it all the more difficult for the economy to be restored to normal functioning once the war was over. The war minister was also afraid that these proposals would destroy the morale of the workers and that in the long run they would be counter-productive.[35]

On 16 September the war minister, Wild, chaired a meeting to discuss the OHL's proposals with leading representatives of industry, among whom were Duisberg, Rathenau and Borsig. Colonel Bauer represented the OHL.[36] Wild was subjected to a ferocious attack from the industrialists, who complained bitterly of the red tape and inefficiency of the war ministry's agencies. They denounced his attempts to secure a modicum of cooperation with the unions as unnecessary and dangerous social experimentation. Industry saw the proposed auxiliary labor law as a means to control their own affairs through the intermediary of a sympathetic OHL. Small and inefficient firms could be destroyed and contracts awarded exclusively to large and efficient companies. A hard line could be taken against labor, profits increased and the competition eliminated.

Whilst this complex struggle between the OHL and the war ministry was going on over the economic organisation of the country, which was bound to have profound effects on all aspects of life, a sudden compromise was reached between Hindenburg and Ludendorff and the Chancellor over the creation of a war office (*Kriegsamt*) to control questions of labor and food supplies. The problem was, however, who should control this new office, which was to be headed by General Groener, a man of exceptional organizational talents. Bethmann cunningly argued that any federal institution would have to come under his jurisdiction and therefore could not be placed under the OHL. He suggested that the *Kriegsamt* be placed under the formal jurisdiction of the Prussian war minister, although in practice it would follow the directives of the OHL. Hindenburg agreed to this proposal, provided that Wild was dismissed and replaced by a man who would toe the OHL's line. This was granted, Wild was replaced by an unimaginative and ultra-conservative officer, General von Stein, and the *Kriegsamt* was created under Groener. The result was disappointing to the OHL. The *Kriegsamt* amounted to little more than a reshuffling of a top-heavy bureaucratic structure and was given ill-defined powers. Groener turned out to be altogether too conciliatory in his attitude to labor for the OHL's taste, and the *Kriegsamt* soon became a political football to be kicked around between the OHL and the war ministry, the Chancellor and the industrialists.[37]

With its partial victory over the war ministry and the creation of the *Kriegsamt*, the OHL now devoted its energies to the auxiliary labor law. The experience of the *Kriegsamt* should have given the OHL ample warning that their idea of placing all Germans between the ages of sixteen and sixty under military control and forcing them to work for the war effort was likely to meet with stiff opposition from those who argued in favor of conciliation and cooperation rather than force in labor relations. But they were determined to push the bill through the Reichstag, not only as a dramatic gesture but also so that if the measure failed the blame could be placed on the politicians.[38] Bethmann argued for watering down the bill to avoid compulsion and to make some concessions to labor. Some industrialists were prepared to accept an amended bill, but others like Hugo Stinnes were horrified at the suggestion that arbitration committees be established in larger factories in which the unions should be represented, insisting that such proposals were seriously detrimental to industry.[39]

The most important debate on the draft bill took place in the Prussian ministry of state in November. Groener argued forcefully that the arbitration committees with union representatives were essential "safety valves" and not a capitulation to militant unionists. The ministers, although many agreed with anti-unionists like Helfferich that this was going too far, finally accepted Groener's argument that token committees were at least better than writing the right to strike into the bill as the unions had asked. Thus the bill was forwarded to the Bundesrat, with the committees still included, on 14 November 1916.[40] Prodded by the OHL, the bill passed the Bundesrat quickly and was passed by the Reichstag with only the fourteen members of the leftwing of the social democrats voting against the bill. The amended bill was celebrated by the social democrats as a triumph and a major step towards war socialism. The Chancellor, Groener, the Reichstag majority, and the unions all hoped, in their different ways, that the bill would mark a new phase in cooperation between capital and labor: both would benefit to the common good of the country and its war effort. Leftwing socialists denounced the bill for what it was: an attempt to curtail the freedom and rights of the working class by binding workers to employers unless they could show that a new job offered a "suitable improvement of working conditions".

The committees were designed to promote understanding between management and labor and to bring workers' complaints to the attention of the employers in any business employing more than fifty workers. They were without any real power or influence, although some able agitators were able to use them as a useful platform. The OHL and their industrialist supporters were quick to blame all Germany's economic ills on the law as amended. They denounced the Reichstag majority for delivering up the country to a rapacious and unpatriotic class, and they blamed the bill for shortages, inflation, higher wages, industrial unrest and any setback in the economy. The government had managed to avoid control of profits, which some Reichstag members had wanted, to the considerable alarm of the industrialists; on the other hand, the OHL had not got the wage controls they initially intended. The auxiliary labor law was certainly not a "triumph of labor", nor was it a triumph for the Reichstag parties. The final bill as amended by the Reichstag differed little from the version that Groener had managed to push through the ministry of state. The social democrats, left-liberals and centre had been called upon to legitimise a measure initiated by the OHL. If, in the history of German parliamentary democracy, this was the beginning of the coalition that was to vote for the "peace resolution" of 1917 and later was to form the "Weimar coalition", the Reichstag majority had also to bear the blame for the failure of the government's policy. It earned the hatred of the OHL and its political allies and got nothing concrete in return.[41]

The OHL's bitter denunciation of the selfishness of the working class was misplaced. It is true that there had been a significant increase in nominal wages during the war. At a rough estimate wages had risen 2.5 times in war industries, and sometimes as much as 1.8 times in other industries. Women's wages, which had started from a much lower base, had risen even more sharply.[42] But these increases were far less than increases in the cost of living, which had risen by about 300% by 1918 and by over 400% by 1919. Nominal wages constantly lagged far behind these increases; only a mere handful of workers were able to keep up with rising prices. In war industries, real wages fell by about 23% during the war and in other industries, about 44%. The talk of highly paid munitions workers, profiting like their employers from the war, is a myth. The common experience of the working class during the war was one of increasing deprivation in spite of rising nominal wages, and the attempt to turn the less fortunate workers against those whose wages had risen somewhat more was a miserable failure. Indeed, many of the leaders of the revolutionary movement in 1918 were from the ranks of the higher paid workers, and there was a marked growth of working class consciousness and solidarity in spite of these wage differences.[43] The acute shortages of the war followed a period in which living standards had risen, and were thus even more painful. The mounting resentment of the working class was directed at the war profiteers, the prosperous and the established. Even the best paid workers were undernourished and overworked, sharing a common experience with their less fortunate fellow workers. The low paid workers therefore were willing to accept the leadership of the better paid in the great strikes of 1917 and 1918.

The growth of working class radicalism during the war was echoed by the increasing militancy of white collar groups, which were also tempted to try industrial action to counteract the fall in their living standards and their loss of social status. Yet the radicalism was hardly reflected in the union leadership. Having long since abandoned any idea of a socialist revolution, many union leaders saw "war socialism" as a significant reform and were determined to use the new institutions to gain some influence over economic planning. For these unionists, and for many of the SPD leaders, the *Burgfrieden* marked the beginning of a new era of cooperation between capital and labor. They hoped that the "war economy" (*Kriegswirtschaft*) was more than an attempt to organise the economy to win the war, and that many of the wartime institutions would remain once the war was over.⁴⁴ The vast majority of employers were determined to oppose any such moves and denounced all concessions to the unions, however trivial and pointless, as the thin end of the wedge or an outright attack on the rightful prerogatives of ownership and control of industry. The OHL was bombarded with complaints from heavy industry about the arbitration committees of the auxiliary labor law, which were held largely responsible for the difficulties of the "turnip winter" of 1916/17.⁴⁵ The strikes of 1917 were used as evidence that far too many concessions had been made to the unions. The OHL now supported the view of many leading industrialists that the auxiliary labor law should be either abrogated or drastically changed and that coercive measures against the working class were essential.⁴⁶ The general commission of the trades unions condemned the strikes and denounced any attempt to use the unions for political ends. The SPD leadership equated wartime strikes with treason.⁴⁷ Even Groener was beginning to feel that the OHL might be right after all, and in April 1917 he announced that he regarded strikes as treason and would act against strikers accordingly. The employers were delighted at this announcement and the union leaders were determined to work with Groener against their own radical rank and file. To show their willingness to cooperate, the traditional demonstrations and political strikes on the first of May were cancelled. Bauer summed up the OHL's thinking on the April strikes by blaming the whole affair on the social democrats, "Jewish liberals", and the weak-kneed Chancellor, Bethmann-Hollweg. These sinister allies, allegedly, had misled a patriotic working class and were plunging Germany headling into a socialist revolution.⁴⁸ By May 1917 Groener had fallen from favor once more for failing to take a firm enough stand against strikers in Silesia. The OHL decided to ignore the *Kriegsamt* and to encourage the commanding generals of the army corps to use their exceptional powers against strikers.⁴⁹ In August Groener was dismissed from his post at the *Kriegsamt* and sent to command a division in the east.

The OHL and a group of industrialists headed by Duisberg continued to demand the abrogation of the auxiliary labor law, which was seen as the root of all the present ills. The employers wanted to be "masters in their own house" and to make no concessions whatever to the unions. Their attacks on the auxiliary labor law became attacks on any form of democratic organization. They demanded a *Gemeinwirtschaft* economy and a society based on the estates (*Standestaat*), part of a reactionary and restorative policy that was designed to establish wartime "robber capitalism" as a norm; they rejected the modern industrial state with its modernizing and rationalistic tendencies.⁵⁰ As the Reichstag would never agree to an abrogation of the auxiliary labor law, for the more the law was attacked by the right the more attractive it appeared to the left, these schemes were hopelessly unrealistic. By October, 1917 the OHL had come to believe that perhaps Groener had been right all along. The union leaders Legien and Gustav Bauer were invited to headquarters. Ludendorff, impressed by the sincerity and reasonableness of the two men, agreed that the employers were partly to blame for the deteriorating situation.[51]

The OHL's conversion to the idea of cooperation with the labor leaders was superficial and uncertain, and under the impact of strikes they returned in the summer of 1918 to the idea of placing workers under direct military control and abrogating the auxiliary labor law.[52] But by this time even the most outspoken industrialists, including Stinnes, Thyssen and Duisberg, realized that the abrogation of the law was impossible and agreed to talk with union leaders. Partly because of a mistaken impression of the military situation after the German offensive of 1918, men like Stinnes and Vogler were pointing out the similarity of interests between employers and employees in that both would share the spoils of a victorious war.[53] After the failure of the German troops in August of 1918, and as the entire political system seemed in danger of collapse by October, the heavy industrialists were prepared to reach some sort of compromise with labor. When the industrialists agreed to the "social partnership" after the November revolution, including an eight hour day and the break with the "yellow" unions, the union leaders imagined that they had won a significant victory and had gained economic democracy. But the industrialists had merely made a tactical compromise and were biding their time before they struck back.

Complaints about excessive profits in industry were amply justified and were a major cause for the mounting social unrest. Even though industrial production dropped significantly during the war, profits and dividends were higher than ever. Rationalization and investment in new machinery necessitated by labor shortage, cheap credits and outright grants to war industry, compensation to firms that were closed during the war, and the refusal of government authorities to control prices all resulted in windfall profits to industry.[54] The rich felt none of the hardships and privations of the poor. Money could buy anything. Indeed, the fall in production of luxury goods was less than the average for industrial goods.[55] The movement towards monopolization, cartelization and trusts speeded up. Large firms grew at the expense of the small competitors. By means of the "War Committees" and "War Societies", which were state organizations in almost all branches of industry, large entrepreneurs were saved from the risks and uncertainties of the free capitalist market without loss of private profit.[56]

In spite of the undoubted benefits to most sectors of industry, attitudes towards the developmet of a more rigidly controlled economy were uncertain and changed with the circumstances of the moment. Small enterprises hoped that a state-controlled interventionist economy might save them from the ravages of the giants, but they soon found that in practice, in the interest of economic efficiency, the state institutions acted largely to the advantage of big industry. The latter complained about bureaucratic inefficiency and excessive red tape and objected to some actions of the state organizations, but realised that these new structures could be used to its own advantage. As the problems of the postwar economy began to attract the attention of the industrialists they realised the need for state help and the continuation of the cartels and syndicates in order to overcome the probable difficulties of supplies of raw materials and high prices on the international markets.[57] Some, like the shipping magnate Albert Ballin and his Hamburg associates, were determined opponents of the idea of a controlled economy and warned that the economy could not be ordered around on the parade ground, but they were swimming against the stream. State intervention in the workings of the economy was a natural consequence of the development of the economy, and was speeded up by the exigencies of a war of attrition.[58]

The main quesiton remains: in whose interests did state intervention take place? The answer is not difficult to find. Heavy industry profited directly from it, used the institutions formed during the war to withstand the revolutionary movement in 1918-1919, and prepared itself to meet the difficult challenge of the postwar years. In these developments the close relationship between the heavy industrialists and the OHL played a vital role. Tactical complaints about the "nationalization of industry" or "communist economic forms" or "state socialism" and denunciation of tiresome controls, bureaucratic meddling and compulsory syndicalization should not be taken as evidence of opposition by industrialists to the growth of state intervention in the economy.[59]

Rising profits and dividends on the one hand, increasing misery among the working population on the other, were the direct causes of the intensification of class antagonisms during the war. For the OHL the only way to overcome this conflict was to offer the glittering prospect of grandiose war aims that alone would justify the sacrifices and injustices of the moment. The OHL's war aims were the most excessive, accurately reflecting the extremist views of their strong supporters in the Pan German League and the *Vaterlandspartei*, although they were always careful to disguise their demands in terms of "military necessity". Thus their Polish policy failed to take into consideration the situation in Poland or the effects on the alliance with Austria-Hungary and was consequently a disastrous failure. Their demands for the domination of Belgium and for the deportation of Belgian workers torpedoed any peace initiatives with the Entente and tied the hands of the chancellor.[60] Impatient with the demands of their allies, the OHL called for an eventual *Anschluss* with Austria-Hungary, by force if necessary, and the creation of a chain of client states in the Balkans under close German control. Similarly, in the Baltic states, the OHL called for direct German domination and rejected out of hand the more flexible and indirect approach of the civilians.[61] At the negotiations at Brest-Litovsk the OHL demanded the most excessive terms, once again excused by military imperatives.[62] The OHL also tried to place the Ukraine under strict military control to exploit to the full the economic resources of the country, but the food supplies taken from the Ukraine were about sufficient to feed the half-million German troops stationed in the country, troops tied down in the east who could have been used on the western front.[63] But the German advance in the east did not stop in the Ukraine. Troops marched to the Crimea, the Black Sea coast and on to the Caucasus. In the north an expeditionary force was sent to help Mannerheim's army in Finland. These adventures brought no military advantages, placed an intolerable strain on German relations with Austria-Hungary and Turkey, and were a constant drain on dwindling reserves of men and material.[64] Even after the major setback on the western front in August 1918 the OHL continued with its annexationist schemes in the east. Only the collapse of the army in the west brought them to an end, although dreams of a German empire in the east lived on to become a main component of Nazi policy.

The constant quarrels between the OHL and the civilian government over war aims and foreign policy have been seen in terms of a struggle between wicked militarists and fundamentally decent and realistic politicians who were trapped in a situation that they could not control.[65] In fact there was a broad similarity of aims on both sides, and the disagreements between them, however passionate, were largely over tactics. Both sides agreed that the war was being fought so that Germany might become a world power with significant territorial additions. The OHL in almost every case demanded outright annexation, direct military rule and the ruthless exploitation of the resources of any territory that came under German control. The civilians favoured a more subtle approach, often arguing for indirect control by means of subservient regimes, elaborate trade agreements and treaty obligations.[66] The OHL was able to support their arguments with spurious military reasoning, and in the last resort Hindenburg and Ludendorff would threaten to resign if their ideas were not supported. Such was their power and prestige that these tactics worked.

until the bitter end, when, as a result of the defeat of the German troops in the west, their views no longer commanded automatic respect.

The exceptional power and influence of the OHL under Hindenburg and Ludendorff was in part due to the unique position of the army within German society, which further strengthened as a natural consequence of the war. But their position was also qualitatively new, and resulted from the structural peculiarities of the German political system. Bismarck had been determined to preserve the social status quo without rejecting economic modernization. As a result opposition groups could not be institutionalized and integrated into the political system but were denounced as "enemies of the Reich". The "loyal opposition" is a particular strength of bourgeois democracy: it provides an important safety valve and encourages participation. Without this system a modern industrial society in its capitalist form is liable to become chronically divided and increasingly ineffective. In the Bismarckian system integration had to be provided by a figure who commanded the respect, approval and admiration of the vast mass of the people. In a conservative and anti-parliamentary system only the Kaiser could serve this function.

The eclipse of the Kaiser's power and authority during the war enabled Hindenburg and Ludendorff to consolidate their power. The decisive step in this direction had been taken on 1 August 1914 when Moltke, then chief of the general staff, had refused the Kaiser's order to halt the German advance into Belgium. From that moment on the imposing title of "Supreme Warlord" was an empty honour. The Kaiser's position was further undermined by the massive campaign to secure Hindenburg's appointment, which gave the new OHL a plebiscitary legitimation that further enhanced their power. It was the Kaiser's constitutional duty to mediate between the civilians and the military, and as the tensions between them grew this function became all the more important. Yet the Kaiser was unwilling to fulfill it. Although he resented the encroachments made by the military on his royal prerogative and had to put up with intolerable behavior from Hindenburg and Ludendorff, he was unwilling to side with the civilians, in part through fear of appearing weak, conciliatory and unmilitary. Thus Hindenburg, a popular hero, widely respected and inspiring general confidence, became a kind of *ersatz* Kaiser, his function to integrate a divided society by the power of his personality and by virtue of his position as chief of the general staff.

This overt attempt to mobilize public opinion behind the army leadership for distinctly political purposes was something new in the history of the German army. The first steps had been taken when the army actively supported the Army League (*Wehrverein*) in 1913, having at first been highly critical of its outspoken behaviour.[67] The army continued to insist that it was unpolitical and concerned solely with technical military matters, but increasingly it used its wide popular support to pursue its political aims. With the formation of the "Fatherland Party" (*Deutsche Vaterlandspartei*) in the summer of 1917 the OHL had a political party which gave it unconditional support.[68] Headed by Wolfgang Kapp and Admiral Tirpitz, the new party was a broad coalition of the extreme right with 1.25 million members within the first year. It called for extreme war aims, an all-out war effort, and an end to any suggestions of political and social reform. The *Vaterlandspartei* was a new and radical form of *Sammlungspolitik*, but its ideology is a striking precursor of later National Socialism. It is more than mere coincidence that the founder of the National Socialist Party (NSDAP), Anton Drexler, was an official of the *Vaterlandspartei*.

The *Vaterlandspartei* was formed as a direct response to the peace resolution of the Reichstag of July 1917; the tabling of the law on the entailed estates (*Fidei Commiss* bill) in January 1917, in which the Reichstag majority had rejected a measure which clearly favoured the landowning aristocracy; and the Kaiser's Easter message in April, which promised to end the three-class suffrage in Prussia when the war was over.[69] The attempts by the Reichstag to increase its influence, which started in 1916 with the discussion of food policy, the formation of a "Main Committee" (*Hauptausschuss des Reichstages*), the revisions of the Auxiliary Labor Law, and the formation of a constitutional committee in the following year, are taken by many historians as evidence of a steady and significant parliamentarization of German political life during the war.[70] The *Fidei Commiss* debate, the Easter message, the role of the Reichstag in the dismissal of Bethmann-Hollweg in July 1917 and of his successor Michaelis in October, the appointment of Payer as Vice-Chancellor to Hertling, are all seen as a continuation of this trend. Victory was then achieved with the OHL's agreement to a parliamentary regime under Max von Baden on 28 September 1918.

The undoubted setbacks over the auxiliary labor bill and the growing rumblings in the Reichstag prompted some rightwing elements to think seriously of a military dictatorship. Carl Duisberg organized meetings of like-minded people from industry and the political right who met at the Hotel Adlon in Berlin to discuss the possibility of a military dictatorship, for by the end of January 1917 Duisberg was convinced that this was the only solution to Germany's problems. The resulting "Adlon Action" was actively supported by the OHL's propaganda machine.[71] At headquarters it was Colonel Bauer who most actively supported the idea of an open military dictatorship. There were times when Hindenburg and Ludendorff were tempted by Duisberg's schemes, but they were reluctant to overthrow the existing political system which offered them so many advantages. The army traditionally had preferred to hide behind the civilians so that, as Ludendorff put it, the chancellor could act as a lightning conductor. The generals knew that their exceptional position was due in large part to their immense popularity, and they did not wish to risk this popularity in playing too prominent a political role. The failures of German policy, in the formulation of which the OHL played a key role, could be blamed on the civilians. The OHL could exercise almost complete control on many aspects of foreign and domestic policy, but it did not have to bear any of the responsibility when those policies failed. Even Bauer could reconcile his views that "to govern means to dominate" with the lightning conductor theory. He comforted himself that at least if things got out of hand the army still controlled the machine guns.[72]

The Adlon Action misfired, in large part because its clandestine actions were skillfully leaked to the press and it soon became all too obvious that Duisberg and his group could not count on significant popular support.[73] Members of the Adlon group then devoted all their energies to the organization of the *Vaterlandspartei.* Again the army cooperated with this venture, using the army educational service of the "Patriotic Instruction" to serve up the propaganda material of their political allies. But the decision to influence the course of political life by means of a parliamentary party rather than directly through an open military dictatorship should not be seen as a a major victory for the Reichstag and for parliamentary democracy. There can be no doubt that the OHL and its political allies were concerned about the growth of democratic forces within the country, which in turn was the direct consequence of the deepening economic, political and social crisis and the growing disenchantment with the war itself. Yet the OHL tolerated the civilian government, and even the Reichstag parties, because they wished them to bear responsibility for any failures in their planning. To this extent the increasing influence of the Reichstag was little more than the illusion of power until the Reichstag and the democratic forces were to be blamed for the failure of the army to win the war.

Ultimately the power of the OHL rested on its military success. With the failure of the Michael offensive of July 1918 Hindenburg and Ludendorff found it increasingly difficult to use their "purely military" arguments with any degree of conviction. On 8 August the OHL had to admit that the situation on the western front was hopeless and that the war would have to be ended. As the power of the OHL began to erode there was fierce in-fighting among upper echelons of the army between those who wanted a quick armistice, those who wanted to continue the war under new leadership, and those who supported Ludendorff's increasingly unrealistic view that the central powers could maintain a war of attrition.[74] Yet even while the foundations of its power seemed to be seriously undermined the army continued with its policy of the "lightning conductor" and found willing accomplices among the civilians. Plans for world power had to be abandoned, while some feared the complete destruction of Germany by the Entente. The most serious consideration of all was how to preserve the social status quo even in defeat and humiliation. Dramatic successes had enabled the old elites to remain in power; failure seemed likely to spell their ruin.

The answer was yet another "revolution from above". Concessions would have to be made as in 1806, 1848, 1867 and 1871 so that the elites could ride the storm and strengthen their position once the crisis was over. Under the able leadership of Admiral von Hintze, the OHL's replacement for Kuhlmann as secretary of state for foreign affairs, the OHL cynically agreed to a programme of parliamentary reform. Agreement over the "Hintze action" was reached at OHL headquarters at Spa on 29 September, and two days later Ludendorff told the section heads of the OHL that as the socialists were responsible for the defeat of the army they were to be brought into the government and thus given full responsibility for the armistice.[75] General Groener pointed out that the parliamentarization was necessary in order to place the odium for defeat at the door of the socialists, and that without a "revolution from above" there could very well be a "revolution from below".[76]

The "Hintze action" was a triumphant success. Pressure for parliamentarization was carefully controlled and directed by the group around the state secretary. The Reichstag played an astonishingly passive role in circumstances where they could easily have gained the initiative. The majority parties would not agree to put forward a candidate for the post of Chancellor and meekly accepted Prince Max von Baden, the man most acceptable to Hindenburg and Ludendorff. By 7 October Ludendorff was confident that it would not be long before the old elites would be back in the saddle again and ruling in the accustomed manner.

The creation of a constitutional monarchy under Prince Max of Baden was thus carefully planned and controlled by the OHL and the foreign office, and from the very outset the parliamentary regime was systematically and constantly blamed for the defeat of the German army at the front. The "lightning conductor" had been transformed into the ideologically active "stab in the back" weapon which was to play such a lethal role in the politics of the Weimar republic. The Reichstag had gained a certain amount of power, but its authority was tainted with the responsibility for defeat, for signing the Armistice, and later for the Treaty of Versailles. Parliamentary democracy was born of defeat and humiliation and the OHL was its midwife. The Reich held together through the turmoil of the winter of 1918-19. The amazing degree of continuity of the election results of 1919 and those of 1912 show how successful the "Hintze action" had been in seizing the initiative from the Reichstag and deflecting the revolutionary tensions into safe channels.[77]

In the immensely complex situation of a seriously divided society under severe pressure, there was an understandable tendency to simplify explanations to the point of primitive sloganeering. Thus, economic problems were blamed on the shortcomings of the Auxiliary Labor Law and the excessive rights of the Reichstag. The desire for peace was due to the unpatriotic schemings of Jewish internationalists and socialist agitators. The mounting social crisis was the result of vulgar profiteers and excessively well paid munitions workers. Legitimate requests for long overdue political reforms were regarded as revolutionary attacks on property. Uncertainty at the course of German politics, dissatisfaction with tiresome government regulations, frustration at the military situation, and the growth of state intervention in the workings of the economy have led historians to talk of a perceptible strengthening of the state apparatus during the war, which left it relatively independent from its previously close association with the old elites, particularly its connections with industry.[78]

This idea, which in its most extreme form suggests that the industrialists feared their loss of control over the decision making process to such an extent that they saw the state as an even greater threat than the militant workers, is a direct response to the historians of the (East) German Democratic Republic, who, true to the theory of "state monopoly capitalism", argue the exact reverse.[79] In the present state of the debate the theoretical weaknesses of both sides are all too apparent. On the one hand, the theory of state monopoly capitalism, although containing many fruitful suggestions, is still at a very rudimentary stage and needs rigorous testing to rid itself of Stalinist notions of the state as the mere executive organ of the monopolies.[80] On the other hand, notions of the "independence" of the state, based largely on the writings of Max Weber, tend either to identify the state with the government or to regard it as some ill-defined and idealistic entity which claims and receives certain rights over the citizenry.[81] Seen in these terms it is clear that the mediating role of the state increased in the war, but this function was performed in close consultation with the industrialists. They in turn had to make concessions and alliances which in a less critical situation they would never have considered. Much the same is true of the army. The OHL under Hindenburg and Groener decided to brave the storm, resisting the advice of the hotheads who were prepared to precipitate a civil war or longed for the futile heroics of a last-ditch stand on the western front and a *levee en masse*. As a result of skilful compromise and hard bargaining, the army resisted all attempts at democratization and remained a potent and reactionary force throughout the Weimar republic.

Even in defeat the army was not discredited. The republicans who had had the republic forced upon them by the OHL had to bear the full blame for defeat. The collapse of the German empire did not lead to the regeneration of German society. The traditional elites were able to defend their power bases in politics, the economy and society at large. The social, economic, political and even psychological structures of Bismarck's Reich proved to be remarkably resilient, surviving in modified forms the experience of defeat and revolution. The refusal to tackle the fundamental problems of this chronically unevenly developed country meant that the fundamental contradictions that had plagued the Reich lived on in new and intensified forms. The ensuing attempt to overcome a chronic crisis plunged Germany into the horrors of Nazi dictatorship.

It is in this context that the persisting question of the continuity of German history must be examined. Discussions of the similarities between Bismarck, Ludendorff, Stresemann and Hitler are liable to be fruitless, for they start from false premises. Outmoded structures survived, and as the process of social change continued, became increasingly anachronistic, menaced and antagonistic. At another level, similarity of aims should not be confused with the forms of domination. Similarities between the ideas and policies of the "silent dictatorship" of the OHL under Hindenburg and Ludendorff and Hitler's "Third Reich" are obvious, but the means chosen to implement these policies were different. The crisis of German society in 1933 was even more severe than that of 1916, and the remedy applied was infinitely more brutal and drastic. That this occurred was due in large part to the failure of the democratic and reforming forces within German society to use the wartime crisis to their own advantage. The relative success of the OHL in preserving the position of the traditional elites and blocking meaningful reforms was to have terrible consequences. The celebration of the triumph of the Reichstag, the unions and the social democrats was tragically premature in 1919. Its apotheosis in historical writing is a cruel mockery.

Notes

1. For a provocative discussion of the concept of "crisis" see Jurgen Habermas, *Legitimation Crisis*, Boston 1975, pp. 1-31.
2. This is a central thesis of Fritz Fischer's two major works: *Griff nach der Weltmacht*, Dusseldorf 1961; *Krieg der Illusionen*, Dusseldorf 1969. Shortened English translations are also available: *Germany's Aims in the First World War*, 1967, and *War of Illusion*, 1973.
3. The social history of Germany in the First World War has yet to be given the full treatment it deserves. A. von Mendelssohn-Bartholdy, *The War and German Society*, New Haven 1937, is a pioneering study full of stimulating insights. Jurgen Kocka, *Klassengesellschaft im Krieg 1914-1918*, Gottingen 1973, is a clever essay based on a curious methodological mixture of Marx and Weber which has still not provoked the discussion which it clearly intended. In more general terms Marc Ferro, *The Great War 1914-1918*, London 1973, is suggestive.
4. Martin Kitchen, *The Silent Dictatorship: The Politics of the German High Command under Hindenburg and Ludendorff, 1916-1918*, London 1976. For a spirited contrary view see Wilhelm Deist, *Militar und Innenpolitik im Weltkrieg 1914-1918*, Dusseldorf 1970, pp. LXIV-LXVI.
5. Max Weber, *Gesammelte politische Schriften*, Tubingen 1958, p. 19.
6. Thorstein Veblen, *Imperial Germany and the Industrial Revolution*, Ann Arbor 1966 (First published 1915).
7. In the "Forward" to the first edition of *Das Kapital*, Marx-Engels, *Werke*, Berlin 1969, vol 23, p. 15. A comparative study of this phenomenon of the transition to modern industrial society, with the old agrarian elites retaining decisive political power, is provided by Barrington Moore Jr., *Social Origins of Dictatorship and Democracy*, Boston 1966.
8. There is a dearth of studies of the industrialisation of Germany. More recent studies include: W. O. Henderson, *The Rise of German Industrial Power*, Berkeley 1975; Helmut Bohme, *Prolegomena zu einer Sozial-und Wirtschaftsgeschichte Deutschlands im 19. und 20. Jahrhundert*, Frankfurt 1968; Knut Borchardt, *The Industrial Revolution in Germany*, London 1972; Wolfram Fischer, *Wirtschaft und Gesellschaft im Zeitalter der Industrialisierung*, Gottingen 1972; F. Lutge, *Deutsche Sozial-und Wirtschaftsgeschichte*, Berlin 1966; H. Mottek, H. Blumberg, H. Wutzmer, W. Becker, *Studien zur Geschichte der industriellen Revolution in Deutschland*, Berlin 1960; Hans Mottek, *Wirtschaftsgeschichte Deutschlands*, Vol. I, Berlin 1959, Vol. II, Berlin 1964; Wilhelm Treue, *Wirtschaftsgeschichte der Neuzeit*, Stuttgart 1966.
9. R. von Friesen, *Erinnerungen aus meinen Leben*, vol. III, Dresden 1910, p. 11. On Bonapartism see E. Engelberg, "Zur Entstehung und historischen Stellung des preussisch-deutschen Bonapartismus", in *Beitrage zum neuen Geschichtsbild*, ed. F. Klein, J. Streisand, Berlin 1956; H. Gollwitzer, "Der Casarismus Napoleons III im Widerall der offentlichen Meinung Deutschlands", *Historische Zeitschrift*, 173, 1952. For a contrary view see A. Rein, *Die Revolution in der Politik Bismarcks*, Gottingen 1957.
10. H. Rosenberg, *Grosse Depression der Bismarckzeit*, Berlin 1967; K. W. Hardach, *Die Bedeutung wirtschaftlicher Faktoren bei der Wiedereinfuhrung der Eisen- und Getreidezolle in Deutschland*, Berlin 1967; H.-H. Herlemann, "Vom Ursprung des deutschen Agrarprotektionismus", in *Agrarwirtschaft und Agrarpolitik*, ed. E. Gerhardt, P. Kuhlmann, Cologne 1969.
11. The useful notion of "negative integration" comes from Wolfgang Sauer, "Das Problem des deutschen Nationalstaates", *Politische Vierteljahrsschrift*, Band 3, 1962.
12. For a detailed study see Dirk Stegmann, *Die Erben Bismarcks*, Cologne 1970.

13. This theme is developed by H.-U. Wehler, *Bismarck und der Imperialismus*, Cologne 1969, and in his essay "Der Aufstieg des Organisierten Kapitalismus und Interventionsstaates in Deutschland" in H. A. Winkler (ed.); *Organisierter Kapitalismus*, Gottingen 1974. The concept of "organised capitalism", which originates from Hilferding, is a very vague theory of limited heuristic value but one which has attracted many West German historians in their search for an alternative theory to the undeveloped and problematic notion of "state monopoly capitalism".
15. P.-C. Witt, *Die Finanzpolitik des Deutschen Reichs von 1903 bis 1913*, Hamburg 1970, for a detailed discussion of this point.
16. Gordon Craig, *The Politics of the Prussian Army 1640-1945*, Oxford 1955; K. Demeter, *Das Deutsche Offizierkorps in Gesellschaft und Staat, 1650-1945*; Manfred Meserschmidt, "Die Armee in Staat und Gesellschaft 1866-1870" and Wilhelm Deist, "Die Armee in Staat und Gesellschaft 1890-1914" in M. Sturmer (ed.), *Das Kaiserliche Deutschland*, Dusseldorf 1970; M. Kitchen, *The German Officer Corps 1890-1914*, Oxford 1968; E. Kehr, *Schlachtflottenbau und Parteipolitik 1894-1901*, Berlin 1930; V. Berghahn, *Der Tirpitz Plan*, Dusseldorf 1971; H. Schottelius and W. Deist (eds.), *Marine und Marinepolitik im kaiserlichen Deutschland, 1871-1914*, Dusseldorf 1972.
17. The phrase is from Fritz Fischer, *Illusionen*, p. 13.
18. K.-H. Janssen, "Der Wechsel in der OHL 1916", *Vierteljahrshefte fur Zeitgeschichte*, 7, 1959; G. Ritter, *Staatskunst und Kriegshandwerk*, vol. III, pp.216-249; Kitchen, *Silent Dictatorship*, pp. 25-45.
19. M. Kitchen, "Hindenburg", in Peter Dennis and Adrian Preston (eds.), *Soldiers and Statesmen*, London 1976.
20. *Spartakusbriefe*, Berlin 1958, p. 273. Werner Basler, *Deutschlands Annexionspolitik in Polen und in Baltikum 1914-1918*, Berlin 1962; Werner Conze, *Polnische Nation und deutsche Politik im ersten Weltkrieg*, Cologne 1958; Paul Roth, *Die Entstehung des polnischen Staates*, Berlin 1926; I. Geiss, *Der polnische Grenzstreifen 1914-1918*, Lubeck and Hamburg, 1960.
21. A. Scherer et, J. Grunewald, *L'allemagne et les problemes de la paix pendant la premiere guerre mondiale*, Vol. I, Paris 1962, p. 492.
22. *Das Werk des Untersuchungsausschusses der Verfassungsgebenden Deutschen Nationalversammlung und des Deutschen Reichstages IV Reihe*, Vol. 2, p. 170; Politisches Archiv des Auswartigen Amts, Bonn Gr. HQ 42, *U-Bootkrieg*, Band 3.
23. Scherer, Grunewald, I, p. 609.
24. These are the figures used in the "Kalkmann memorandum" which was designed to answer the pessimistic figures used by Helfferich to argue against unrestricted submarine warfare. See K. E. Birnbaum, *Peace Moves and U-Boat Warfare*, Stockholm 1958.
25. E. Ludendorff, *Urkunden der Obersten Heeresleitung uber ihrer Tatigkeit 1916/18*, Berlin 1920, p. 322.
26. Georg Alexander von Muller, *Regierte der Kaiser? Kriegstagebucher*, ed. W. Gorlitz, Gottingen 1959, 6-2-17.
27. Z. A. Zeman, *Germany and the Revolution in Russia 1915-1918*, London 1950; W. B. Scharlau and Z. A. Zeman, *Freibeuter der Revolution. Parvus Helphand*, Cologne 1964.
28. Leo Stern (ed.) *Archivalische Forschungen zur Geschichte der deutschen Arbeiterbewegung*, Band 4/I-IV, Berlin 1959; J. Petzold (ed.), *Deutschland im Ersten Weltkrieg*, Vol. III, Berlin 1969.
29. Walter Rathenau, *Politische Briefe*, Dresden 1929, no. 40; Erich Ludendorff, *Meine Kriegserinnerungen 1914 bis 1918*, Berlin 1919, p. 216; Bundesarchiv Koblenz, Nachlass Bauer, Band 11.

30. Friedrich Zunkel, *Industrie und Staatssozialismus*, Dusseldorf 1974, pp. 32-41.
31. Ludendorff, *Urkundun der OHL*, p. 63.
32. BA Koblenz Bauer, Band 2. The slightly amended final version is in *Urkunden der OHL*, p. 65.
33. The best known works in this vein are: Richard von Moellendorf, *Deutsche Gemeinwirtschaft*, Berlin 1916; also his *Von Einst zu Einst. Der Alte Fritz, J.G. Fichte, Freiherr vom Stein, Friedrich List, Furst Bismarck, P. de Lagarde uber Deutsche Gemeinwirtschaft*, Jena 1917; Walter Rathenau, *Von kommenden Dingen*, Berlin 1917, and his *Die neue Wirtschaft*, Berlin 1917.
34. Ritter, *Staatskunst und Kriegshandwerk*, vol. 3, p. 423.
35. *Der Weltkrieg*, vol. IX, p. 37.
36. Deist, *Militar und Innenpolitik*, vol. I, p. 486.
37. Kitchen, *Silent Dictatorship*, pp. 72-75.
38. Ludendorff, *Urkunden der OHL*, p. 81.
39. G. D. Feldman, *Army, Industry and Labor in Germany, 1914-1918*, Princeton 1966, p. 204.
40. Ibid., p. 209.
41. "Triumph of labor" is Feldman's phrase: see chapter IV of *Industry and Labor*. The debate on the *Hilfsdienstgesetz* still goes on. Kocka, *Klassengesellschaft*, p. 115 sees it as an important step in the strengthening of the constitutional role of parliament. Wehler, *Das Deutsche Kaiserreich*, p. 205 is more cautious and correctly stresses the satisfaction of the unions and the rightwing SPD as a further stage in the process of "negative integration", adding that the dissident left saw the situation more clearly. East German historians are unanimous in attacking the law as a vicious attack on the working class.
42. For a useful summary and helpful bibliography see Kocka, *Klassengesellschaft*, pp. 12-21.
43. E. Kolb, *Die Arbeiterrate in der deutschen Innenpolitik 1918/19*, Dusseldorf 1962; P. von Oertzen, *Betriebsrate in der Novemberrevolution*, Dusseldorf 1963; W. Tormin, *Zwischen Ratediktatur und sozialer Demokratie*, Dusseldorf 1954. G. D. Feldman, E. Kolb and R. Rurup, "Die Massenbewegungen der Arbeiterschaft in Deutschland am Ende des Ersten Weltkrieges (1917-1920)", *Politische Vierteljahrsschrift* 13, 1972 argue that there was bitter resentment by the poor workers for the "rich" ones. There is however very little evidence for this assertion, and class solidarity was remarkably strong and grew stronger during the course of the war.
44. Much further work needs to be done on the history of the trades unions during the war. Useful material can be found in: H. J. Varain, *Freie Gewerkschaften, Sozialdemokratie und Staat, Die Politik der Generalkommission under der Fuhrung Carl Legiens (1890-1928)*, Dusseldorf 1956; F. Opel, *Der deutsche Metallarbeiterverband wahrend des Ersten Weltkriegs und der Revolution*, Hanover 1957; W. Richter, *Gewerkschaften, Monopolkapital und Staat im Ersten Weltkrieg und in der Novemberrevolution 1914-1918*, Berlin 1959.
45. BA Koblenz, Nachlass Bauer, Band 14. Memo. from the Dortmund coalowners. DZA Potsdam Reichskanzlei, Notstanden 2430, Krupp to Chancellor 26-2-17.
46. DZA Potsdam Reichskanzlei, Allgemeines 2398/10.
47. Hans Herzfeld, *Die deutsche Sozialdemokratie und die Auflosung der nationalen Einheitsfront im Weltkrieg*, Leipzig 1928, p. 317.
48. BA Koblenz, Nachlass Bauer, Band 2.
49. Ludendorff, *Urkunden der OHL*, p. 186.
50. Zunkel, *Industrie und Staatssozialismus*, p. 112.

51. Ludendorff, *Urkunden*, p. 94.
52. PA Bonn, Weltkrieg geheim 40, Hindenburg to Hertling 18-6-1918.
53. Zunkel, *Industrie und Staatssozialismus*, p. 133.
54. For a useful chart see Kocka, *Klassengesellschaft*, p. 26.
55. R. Wagenfuhr, *Die Industriewirtschaft* (Vierteljahreshefte zur Konjunkturforschung, Sonderheft 31), Berlin 1933, p. 23.
56. L. Grebler and W. Winkler, *The Cost of the World War to Germany and to Austria-Hungary*, New Haven 1940, p. 39.
57. Zunkel, *Industrie und Staatssozialismus*, pp. 172-200 for further discussion.
58. H. A. Winkler (ed.), *Organisierte Kapitalismus*, article by H.-U. Wehler, pp. 36-57.
59. Kocka, *Klassengesellschaft*, p. 117 greatly exaggerates this point and makes it a major theme of his book.
60. Willibald Gutsche, "Zu einige Frage der staatsmonopolistischen Verflechtung in den ersten Kriegsjahren am Beispiel der Ausplunderung der belgischen Industrie und der Zwangsdeportation von Belgien", in *Politik im Krieg 1914-1918*, ed. Fritz Klein, Berlin 1964; Fernand Passelecq, *Deportation et travail force des ouvriers et de la population civile de la Belgique occupe*, Paris and New Haven 1928.
61. Kitchen, *Silent Dictatorship*, pp. 211-216.
62. W. Hahlweg, *Der Diktatfrieden von Brest Litovsk 1918 und die bolshevistische Weltrevolution*, Munster 1960; *Deutsch-Sowjetische Beziehungen von den Verhandlungen in Brest Litovsk bis zum Abschluss des Rapallovertrages*, Berlin 1967.
63. Peter Borowsky, *Deutsche Ukrainepolitik 1918*, Hamburg 1970; John S. Reshetar, *The Ukrainian Revolution 1917-1918, A Study in Nationalism*, Princeton 1952.
64. See particularly W. Baumgart, *Deutsche Ostpolitik 1918*, Munich 1966; C. Jay Smith, *Finland and the Russian Revolution 1917-1922*, Athens, Georgia 1958; W. Hubatsch, "Finnland in der deutschen Ostseepolitik 1917/18", *Ostdeutsche Wissenschaft*, Band II, 1955; General Graf Rudiger von der Goltz, *Meine Sendung in Finnland und im Baltikum*, Leipzig 1920.
65. This is the view of Gerhard Ritter and his supporters.
66. These problems are analysed in meticulous detail by Fritz Fischer in *Griff nach der Weltmacht*.
67. Kitchen, *German Officer Corps*, pp. 136-139.
68. Stegmann, *Die Erben Bismarcks*, pp. 497-519.
69. Kitchen, *Silent Dictatorship*, pp. 127-138.
70. Kocka, *Klassengesellschaft*, p. 108; J. V. Bredt, *Der Deutsche Reichstag im Weltkrieg*, Berlin 1926; Dieter Grosser, *Vom monarchischen Konstitutionalismus zur parlamentarischen Demokratie. Die Verfassungspolitik der deutschen Parteien im letzten Jahrzehnt des Kaiserreiches*, Hague 1970.
71. DZA Potsdam Reichsamt des Innern, 1963.
72. For Bauer's political views see Martin Kitchen, "Militarism and the Development of Fascist Ideology: The Political Ideas of Colonel Max Bauer, 1916-1918", *Central European History*, vol. VIII, no. 2, Sept. 1975.
73. Conrad Haussmann, *Schlaglichter. Reichstagsbriefe und Aufziechnungen*, Frankfurt 1924, p. 103.
74. Kitchen, *Silent Dictatorship*, pp. 247-267.
75. PA Bonn AA Weltkrieg 23 Geheim, Band 32.
76. Wilhelm Groener, *Lebenserinnerungen. Jugend, Generalstab, Weltkrieg*, Friedrich Freiherr von Gaertringen (ed.), Gottingen 1957, p. 466.
77. P. Mott, *Der Reichstag vor der improvisierten Revolution*, Cologne 1963, p. 357.

78. This thesis is central to Kocka's book; see also Zunkel, *Industrie und Staatssozialismus*, p. 14.
79. Kocka, *Klassengesellschaft*, p. 117; G. Feldman, "German Big Business Between War and Revolution: The Origin of the Stinnes-Legien Agreement", in G. A. Ritter (ed.), *Entstehung und Wandel der modernen Gesellschaft. Festschrift Hans Rosenberg*, Berlin 1970, pp. 312-341.
80. R. Gundel, *Zur Theorie des staatsmonopolistichen Kapitalismus*, Berlin 1967; Paul Boccara, *Etudes sur le capitalisme monopoliste d'etat: sa crise et son issue*, Paris 1973; S. Richter and R. Sonnemann, "ZurProblematik des Ubergangs vom vormonopolistischen Kapitalismus zum Imperialismus in Deutschland", *Jahrbuch fur Wirtschaftsgeschichte*, II, 1963; R. Sonnemann and S. Richter, "Zur Rolle des Staates beim Ubergang vom vormonopolistichen Kapitalismus zum Imperialismus in Deutschland", *Jahrbuch fur Wirtschaftsgeschichte*, II and III, 1964; *Der Thesenstreit um "Stamokap", Die Dokumente zur Grundsatzdiskussion der Jungsozialisten*, Reinbek 1973.
81. Max Weber, *Wirtschaft und Gesellschaft*, Cologne 1964, p. 39. Karl Marx, *Der achtzehnte Brumaire des Louis Bonaparte*, in Marx-Engels *Werke*, vol. 8, Berlin 1969, p. 198. Friedrich Engels, *Der Ursprung der Familie, des Privateigentums und des Staats*, in Marx-Engels *Werke*, vol. 21, Berlin 1969, pp. 164-168.

THE CIVILIAN FACTOR

Mutiny in the Mountains: The Terrace "Incident"

Reginald H. Roy
University of Victoria

In all of Canada, there was probably no army camp of its size as remote from the mainstream of Canadian life as Terrace, British Columbia. Tucked in the mountains beside the Skeena River, its population of about 500 depending largely on the lumbering business, Terrace relied on the Canadian National Railway as its main means of communication with the outside world. Located between the port of Prince Rupert, on the coast, and Prince George, 488 miles inland, this sleepy village was selected in 1942 as the site for one of the infantry brigades rushed to British Columbia following the outbreak of war with Japan. As Japanese troops pushed to the borders of Burma and the Aleutian islands, the fear of a Japanese attack on the coast of British Columbia increased to such an extent that two infantry divisions were sent to the province to repel any such attempt. Prince Rupert, the ocean terminus of the Canadian National Railroad, was looked upon as a potential danger point and was defended by coastal and anti-aircraft batteries and some infantry. A brigade of troops in Terrace could be used as a strategic reserve to counter any attempt by Japanese forces to strike at Prince Rupert and move inland along the line of the railway.

In 1942, therefore, Terrace experienced more activity than it had for decades. Barracks were built for infantry battalions on the outskirts of the village, and shops, quartermaster stores, a drill hall, a hospital, and other buildings were constructed to house and maintain engineer, medical and other support groups. Later, a mountain warfare school was established in Terrace to train troops from Pacific Command. South of the Skeena, several miles from the village, a Royal Canadian Air Force station and airfield were constructed as a base for fighter and bomber reconnaissance squadrons. By late 1943 Terrace was a "garrison village" with the military outnumbering the civilian population by about eight to one.

During the next two years the brigades in Terrace were rotated as a matter of policy. As the threat of a Japanese raid receded, the General Officer Commanding-in-Chief, Pacific Command, Major-General G. R. Pearkes, V.C., had the number of divisions under his command reduced from two to one. By the spring of 1944 the 6th Division had one brigade at Nanaimo on Vancouver Island, another at Vernon in the Okanagan Valley and the third, the 15th Canadian Infantry Brigade, stationed at Terrace. With two brigades on the mainland and one on Vancouver Island, together with some additional garrison troops at key locations close to important seaports, there were ample forces to deal with the decreasing possibility of an enemy attack.

The proportion of conscripts or Home Defense (HD) men in the regiments stationed in British Columbia had always been high in comparison to the number of volunteer or General Service (GS) men. Following the commitment of Canadian troops in Sicily and Italy in the summer of 1943, and especially before and during the invasion of France and the Normandy campaign in 1944, the Department of National Defense urged conscript soldiers to volunteer for overseas service. Within the units stationed in British Columbia, officers did their utmost to convert "HD" men to "GS" status. In newspapers, on the radio, on billboards, from the pulpits and by any other means which public relations firms could devise, there was a constant stream of propaganda aimed at the conscript soldier. The message was always the same: the army fighting overseas needed reinforcements. Since the government had conscripted men only for service at home, it was the duty of the HD man to volunteer, "to go GS" and support his comrades fighting in France or Italy.

The campaign to get the HD soldiers to "go active" continued throughout 1944. At the same time, many GS volunteers in Canada entered the stream of reinforcements for overseas when the Department of National Defense disbanded the 8th Infantry Division in British Columbia and reduced the number of garrison troops in Pacific Command. As a consequence, the regiments remaining in British Columbia gained an ever increasing proportion of conscripts. Home Defense soldiers from disbanded regiments were sent to units in the 6th Division until over 95 per cent of the non-commissioned officers and men in most regiments were conscripts. For years many of them had resolutely refused every appeal to volunteer. Their reasons for not volunteering were as varied as the areas of Canada from which they came, but a great many claimed that if the government wanted or needed them, it would conscript them for overseas service. Since Prime Minister W. L. Mackenzie King had repeatedly declared that his government would not impose conscription for overseas service, it was a refrain which the conscripts could use as a shield for their own personal motives for avoiding service overseas.

By the autumn of 1944 three of these regiments, the Prince Edward Island Highlanders, the Prince Albert Volunteers and Les Fusiliers du St. Laurent, were part of the 15th Canadian Infantry Brigade headquartered in Terrace. Nearby, forming part of the camp, was the 19th Field Ambulance, RCAMC. In all probability, if any soldier in the brigade had been given the opportunity to be posted elsewhere in Pacific Command he would have taken it. Describing the morale of the troops in Terrace, one report stated:

> The average soldier, regardless of the entertainments and pasttimes provided in the camp area, hates Terrace and everything about it. This is due to the lack of entertainment in the village area as the troops are forced to spend their off-hours wandering and congregating in crowds along the small streets and the few public places. This fact is most noticable during the week-ends, Saturday nights especially. [1]

Despite the titles of regiments, the number of men in each unit who had joined them when they mobilized, or who came from the area where they originally recruited, was small. Periodically since 1942, conscripts from all parts of Canada or, later, from disbanded units of the 8th Division came to replace men who volunteered for overseas service. These new drafts, like the conscripts in the units they joined, had been subjected to various persuasions to volunteer. A number of factors contributed to their determination to resist the call to go overseas. There was a general optimism that the war's end seemed only a few months away, and consequently that their services were not really necessary.

At the same time, the appearance of the casualty lists in the newspapers strengthened the apprehension of many who feared that, if they did reach the front-line, their chances of being killed or wounded were by no means remote. Another factor was the ease with which the Home Defense soldier could obtain farm leave, promotions and special courses. Canada was making a mighty war effort, and the amount of food supplies and war equipment she was shipping to Great Britain and to her allies made her economic contributions fourth among all the nations fighting the Germans. The shortage of manpower made it easy to get leave for several months to work on the farm or in the forests. To return to army life and discipline in the autumn, after tasting civilian freedom and the extra pay involved, increased the resentment of many soldiers against the whole concept of the National Resources Mobilization Act, which had given the authority for conscription. Finally, among the "hard core" conscripts, there was a deeply ingrained and stubborn determination not to volunteer no matter what the situation was in Europe.[2] "When asked why they are not GS", one Commanding Officer wrote of the conscripts in his company,

> They state that they do not owe the country anything, that during the "Depression Years", the country didn't care whether they were alive or not, but now it is at war, they are wanted. When it is pointed out that there are plenty of men overseas who were just as badly off as they were, they merely state: "If they want to be suckers, we aren't."
>
> The general GS complaint is why people like these be allowed to live in this country and derive benefits from the efforts of Canadians who are fighting and helping win the war, so that this country may remain free. When the peace has been won, this will be a great country, and why should these spineless jellyfish be allowed to progress with the country, when they have not had the courage to help her in time of trouble.[3]

In mid-November, when the new Minister of National Defense[4] was making a last, desperate appeal to the conscripts to volunteer for overseas service, Major-General R. O. Alexander was conducting his usual inspection tour of troops in Pacific Command. In Terrace he found much room for improvement. In Brigade Headquarters he noted a large turnover in officers and added that most of the former good NCOs, who were volunteers, had left. He was not impressed with the information rooms which Pacific Command had ordered set up to keep the troops up to date on events at home and on the battlefronts. "The troops," he wrote, "were dependent on the radio and newspapers for what is going on anywhere. Reception on the radio is very bad and is very seldom listened to. Receipt of newspapers is spasmodic and three or four days late when they arrive. The result is that there appears to be a complete indifference as to what is going on in the war. . ."[5]

The brigade was not in an efficient state, Alexander felt, adding: "The contributory causes are that the pick of one Battalion is with the Polar Bear Force[6] leaving inferior men in the Unit and a large change over of NCOs which has occurred lately due to GS men being sent overseas."[7] After inspecting the Prince Edward Highlanders he commented that of the normal establishment of about 800 all ranks, 122 were away on command and 64 were on leave. Those remaining were "a mixed lot." Many were reallocated artillerymen, and there were a "large number of men of Central European descent." "There appears to be," he wrote, "an attitude of lassitude and almost defeatism throughout the unit. This may be due in a large measure to the number of officers and NCOs who have left the Regiment lately."[8] The Prince Albert Volunteers, although rated generally more efficient, had the fewest men available of the three units since 436 were "on command." Over 300 were with Polar Bear Force, and what was left was "of mediocre quality." Over three dozen were absent without leave, while others were attending schools, and so forth. He noted, too, that during the autumn, 500 men from the Volunteers had been away on Harvest Leave.[9] Alexander's report on the other units and sub-units in the brigade, although differing in detail, was similar. Regimental esprit de corps was at a low ebb. With the almost complete lack of a Japanese threat to stimulate their training, and their apparent immunity from overseas service, most of the men in the brigade accepted their role in the armed forces with an attitude that ranged from hostile to apathetic.

Shortly after the Inspector General left Terrace, all the senior officers in the brigade were ordered to Vancouver to attend a conference called by the G.O.C.-in-C., Pacific Command, Major-General G. R. Pearkes. Pearkes had just returned from Ottawa where he and other senior Canadian officers had been instructed by the new Minister of National Defense once more to mount a major campaign to convert HD soldiers to GS status. Before Pearkes arrived to address the assembled officers, reporters interviewed a number of them. On the following morning, the Prime Minister was extremely annoyed to read in the newspapers an account which gave the impression that senior officers in Pacific Command believed the government's plan to raise the number of volunteers needed for overseas reinforcements would fail. At the same time the senior army officers at National Defense H.Q. told the Minister of Defense that the only way to obtain the necessary infantry reinforcements was to bring in conscription for overseas. Faced with this crisis, Mackenzie King decided that despite his promises to the contrary, the time had come to conscript men for overseas service. On the following day, November 23, the fateful decision was announced in the House of Commons.[10]

This complete turnabout in government policy surprised everyone. The timing of the announcement was especially awkward to Pacific Command. Immediately following the newspaper account of the so-called "press conference" in Vancouver, Pearkes was told that Lt. Gen. E. W. Sansom was flying to Vancouver to investigate what appeared to be a challenge to government policy. He was instructed, therefore, to reassemble the officers who were at this time now returning to their various camps and units. Some, of course, were close at hand. A ferry trip to Victoria took only five or six hours, and one could reach Vernon in the Okanagan Valley by car in ten hours driving time. The normal route from Terrace, however, was by train via Prince George and Jasper or by train and coastal steamer via Prince Rupert. Usually this would take a day and a half when the trains were running. Pearkes had ordered his officers back to Vancouver when, shortly after noon on November 23, he received a telephone call from Ottawa telling him that the Order-in-Council had been signed conscripting 16,000 Home Defense soldiers.

The suddenness of the news left no time to prepare the troops for the shock. Pearkes, therefore, wired the Adjutant-General two hours later suggesting that in view of "possible repercussions due to this information reaching the troops, I consider it important that Commanding Officers should remain with their units at the present time."[11] He requested a postponement of the investigation until Monday, November 27, to give the senior officers the opportunity to deal with any disturbances that might arise. His advice was not accepted. At the same time, Pearkes sent a top secret telegram to his division and formation commanders informing them about the new policy. At Headquarters, 15th Infantry Brigade in Terrace, the deciphered telegram was received at seven-thirty in the evening, almost four hours after it had been sent.[12] By that time almost every man in the brigade had heard the news which had been broadcast over the radio late that afternoon.

During the evening of 23 November the news about conscription probably formed the major topic of conversation wherever the men gathered. It would appear that in each unit in Terrace, a comparatively small number of men, boiling with resentment over what they considered to be the breaking of the Prime Minister's promise, decided to take action which, in legal terms, was to result in a mutiny. The secrecy in which they acted and the methods they used make it impossible, now, to describe accurately the planning and direction of events from the point of view of the "committee of other ranks" which assumed control of the camp. The events themselves, however, bear witness to the desires of the mutineers.

Acting Captain G. A. Anthony of the Prince Albert Volunteers was in the wet canteen on the evening of 23 November when he saw a number of men from the Fusiliers du St. Laurent talking about the news. It was unusual for the Fusiliers to be there, but there was no trouble. The next morning his company clerk reported that the Fusiliers du St. Laurent had sent men over to organize brigade resistance to the conscription order. Representatives from all units were to meet. The planned organization and demonstration, the clerk continued, was to be considered a threat "followed by threats of physical violence and shooting if necessary."[13] Nothing untoward happened that morning, but the brigade Intelligence Officer received a wire, repeated the following day, from Pacific Command Headquarters ordering him to advise them about the attitude of the men towards conscription and of any incidents which might occur. This officer was to be very busy over the weekend.

After the noon meal on Friday, 24 November, the trouble started. The Acting Commanding Officer of the Fusiliers was approached by this Acting Adjutant, Lt. L. A. Pellerin, who told him that two companies of the unit were in their huts and refused to go on parade. Major H. J. Thuot went to the huts where the men were lying on their beds or lounging around. Thuot talked to them without effect, the men informing him that they wanted it known they were against conscription. Thuot then went to Brigade Headquarters to talk to the Acting Brigade Commander, Major J. S. Wright, who was also Acting Commanding Officer of the P.E.I. Highlanders. Together with the Brigade Major the three officers discussed the situation. "We came to the conclusion," Thuot said later, "that we should not take any measures that would antagonize the men or bring about [a] fracas between officers and themselves. We decided that it was best to impress on the men that we were not afraid of anything; therefore wet canteens were left open and no guards were posted on ammunition dumps."[14]

News of the action by the two companies of Fusiliers spread rapidly. Meanwhile, evidently gathering confidence owing to the lack of any action taken against them, those in the brigade who were controlling the opposition to their officers widened their contacts, began to threaten any who attempted either compromise or obedience to orders, and planned a bolder move. After the evening meal the Fusiliers' Quartermaster, ordered to check the security of his stores, reported all was well. At 2300 hours that night, however, Thuot was told that the doors of one of the storerooms had been opened and arms taken out. Moreover, 50,000 rounds of rifle ammunition had been taken from the ammunition dump together with 15,000 rounds of Sten gun ammunition and four boxes of hand grenades. Thuot immediately ordered his Regimental Police to guard the remaining ammunition but they refused. As a result he had to put some of his officers on guard duty. No one attempted to interfere with this officer's guard either that night or later. Actually, the mutineers had more than enough ammunition for their needs.

Elsewhere in British Columbia on Friday and on Saturday there were demonstrations by conscripts. In Vernon, especially, a large body of troops marched through the town shouting "down with conscription." The men in the demonstration, though noisy, returned to the camp in good order. The only incident occurred when an officer was roughly pushed aside as he tried to remonstrate with the men. This demonstration received wide newspaper and radio coverage, which was magnified out of proportion and presumably further encouraged those in Terrace. The Commanding Officer of the Prince Albert Volunteers stated later that "if news of disturbance, parades, et cetera at other points had been suspended from radio programs", the later demonstrations within the 15th Infantry Brigade would never have taken place.[15]

By Saturday the mutineers were in control. After breakfast virtually no one attended the morning parade. Located next to the Fusiliers barracks was the 19th Canadian Field Ambulance. Captain C. H. Belanger was Transport Officer, Acting Adjutant and Acting Commanding Officer. In this small unit there were only two men who were volunteers, not counting three officers. The men in the medical corps had been warned the previous evening that they would suffer if they went on parade. As a result, on Saturday morning, Belanger cancelled the outdoors parade, assembled his men in the canteen and told them "not to ally themselves with the strikers."[16] In the Fusiliers' lines Major Thuot, en route to and from Brigade Headquarters, reported that he noticed "some of our men going down to the Prince Edward Island Highlanders and the reverse."[17] In the Highlanders lines only the Pipe Band, the Signals Platoon and part of Battalion Headquarters were on parade. Part of "A" Company was not on parade since it had provided guards for some of the ammunition dumps during the night. This company, commanded by Captain B. D. MacGillivary, remained under his influence more than almost any other sub-unit in all three infantry battalions.

At Brigade Headquarters counter measures were put in progress. Early Saturday morning, Lt. Col. W. B. Hendrie, commanding the Mountain Warfare School, visited Headquarters and was informed of the most recent developments. Following the advice of Brigadier Murdock, who was en route to divisional headquarters, Hendrie, as senior officer in the area, decided to assume acting command of the brigade and take action to limit the effect of the mutiny. Hendrie ordered the brigade paymaster to place all money in paymasters' offices in the bank in Terrace. He contracted the local telegraph office and, fearing that the mutineers might be in communication with other military camps, he ordered that no telegrams be sent by military personnel without his permission. He requested the Chief Magistrate in Terrace, Judge Robinson, to close the beer parlors and liquor stores until further notice. Hearing there was a boxcar full of ammunition and another of beer and liquor at the railway siding, he asked the station agent to remove both as quickly as possible. Hendrie also ordered the Acting Commanding Officers to do nothing unless the mutineers molested the citizens of Terrace or private property. Normal administrative fatigue and duties were carried out in the units and no physical action had been taken against the officers or NCOs. Hendrie wanted to avoid any clash in a situation that was critical but not yet dangerous. Aside from mounting a guard on the Camp Supply and Transport Office with trusted men, he took no further steps at this time. Hendrie reported everything he had done to his divisional commander, Major-General H. N. Ganong, and to Pearkes in Vancouver.[18]

The coming and going of small groups of men between the lines of the Fusiliers and the Highlanders, one can assume, was to prepare for the massive demonstration staged early Saturday afternoon. Majors Wright and MacMillan, the acting COs of the Highlanders and the Volunteers, had been forewarned, and Wright had ordered his officers not to interfere nor to wear side arms. The demonstration was a show of strength and, as may have been planned, it had the effect of further intimidating those who wished to remain neutral or, at any rate, stay out of trouble. The parade started in the line of the Fusiliers and came down the hill past Brigade Headquarters. The men, wearing their steel helmets and carrying their arms and respirators, were well organized and marched in good order. Carrying banners reading "Down with Conscription" and similar slogans, they paraded into the Highlanders camp and then through Terrace, yelling, singing and shouting to the Prince Albert Volunteers' camp. There they picked up more demonstrators. "This parade," Lt. Col. Hendrie remarked later, "reminded me rather of an overgrown lot of school boys who had just told the teacher to jump in the lake and she had done it."[19]

The parade lasted two hours. Of the three infantry units, the Volunteers supplied the fewest numbers. Their commanding officer said later:

> I am of the opinion that this unit was intimidated and threatened by other units within the brigade, and had this... not taken place, I am sure this unit would not have taken part in any demonstration.[20]

The demonstrators, after marching around town and through the camps, broke up into their respective units and returned to their lines. An estimated 1,500 men had taken part in the parade and had demonstrated their unity against conscription. It was a force to be reckoned with. Moreover there were widespread rumors that the Fusiliers had mounted anti-tank guns and mortars in their lines, which were situated on high ground overlooking several other camps and the village. This further intimidated those remaining loyal to their officers. After the parade, for example, the men in "A" Company, Prince Edward Island Highlanders, refused to continue guard duty on the ammunition dumps. This duty was continued by officers and senior NCOs. In the 19th Field Ambulance lines, situated next to the Fusiliers, only about two dozen men had taken part in the parade. Much credit was due his unit, Captain Belanger said, "all the more so considering the fact that strikers armed with Stens were roaming through our lines, trying to get our men to join the parade." He continued:

> All evening scores of infantrymen roamed through our men's huts endeavouring to "convert" our men to their course and warning them of reprisals if they did not participate in the strike. This system of intimidation was having some measure of success. We learned that the strikers intended going to church armed, for fear that the officers might attempt to lock them in the Drill Halls [where the church services were held].[21]

Saturday night was quiet and again there were no incidents. Routine work and fatigues were carried out as was to be the case during the following week. In the recreation hall the usual films were shown and the "no smoking" rule was observed. In the Fusiliers' wet canteen the word had gone around that if anyone got drunk he would be placed in detention, presumably on the orders of those controlling the meeting. In the same unit the mutineers grounded all vehicles except for one Jeep and a small truck.

Sunday also started out peacefully with a volunteer church parade. Major Thuot asked the padre of the Fusiliers to announce that he wanted to meet representatives from the various companies in his office at noon. The meeting, held in the padre's office at the request of the company representatives, was not very helpful. Aside from stating that they should not be considered leaders but representatives, the dozen men requested that the unit's wet canteen should be closed to prevent any trouble. When Thuot questioned them regarding their motives, they presented him with a typed, unsigned letter which, they claimed, represented the opinion of the men in the battalion. They wanted both the public and the government in Ottawa to know that they opposed conscription. They wished to hear this publicly stated in the newspapers and on the radio, and wanted a telegram sent directly to the federal authorities. They also wanted to be returned to Quebec with a guarantee of their security until they returned.[22] Except for mentioning Quebec, the letter was very similar to the one given to Major J. S. Wright of the P.E.I. Highlanders on the same day.

On the same afternoon the mutineers turned against their sergeants. Shortly after noon, about 75 Fusiliers entered the Sergeants' Mess and ordered them to take down their stripes and join them within three hours. Having challenged the sergeants to "go active" or join them,[23] the group then went to the unit's Quartermaster stores and took a number of Sten machine carbines and Bren light machine guns. However, the Fusiliers' officers "had taken away most of the breech blocks of the Brens, as well as the firing pins out of the mortars."[24] A similar challenge had been given at the same time to the sergeants in the other units. Major Thuot, after consulting with the Acting Brigade Commander, advised his sergeants to remove their stripes. He hoped that if the sergeants moved from their quarters into the men's huts they would have a moderating influence on the men. Further, and the other unit commanders must have agreed, he wanted to "avoid any fights, that once blood was shed there would be no backing out of mutiny for the men."[25] Thus, when the group returned to the sergeants' quarters later that afternoon, they found no one wearing their rank badges. While these events were taking place, an incident in the Highlanders' camp underlined the loss of authority of even the senior warrant officers. There the regimental sergeant major, having posted the duty roster, found one man refusing to work. The man was ordered by the R.S.M. to the guard room under arrest, but en route there he was taken away from the provosts accompanying him.

The methods being used by the ringleaders were becoming more apparent as time went on. By using the battalion's wireless sets as well as the camp telephone system, the leaders were in constant communication with each other. Moreover, it seems that the leaders in one battalion were using the threat of action from a larger group in another battalion to intimidate anyone who appeared reluctant to join in defying their officers and NCOs. As Major Wright of the Highlanders reported:

> It was becoming apparent from reports I as receiving, from senior NCOs and Officers, that a large majority of the personnel appeared to have been terrorized into their action. This was definitely verified later on during the disturbance. It appeared that those who were ring leaders within our own unit were using the Fusiliers du St. Laurent as a threat against the men who were not with them in our unit. I know that the rumour was current that Bren guns were mounted on the top of the hill and that the men had been told that anyone who appeared on the parade ground would be fired upon.[26]

Equally important to the ringleaders was their desire not to have the men influenced by the officers who attempted, whenever possible, to speak to groups of the men and urge them to return to normal duty and discipline. In many ways the ring leaders had the advantage, since most of the platoon and company officers had only recently been posted and had not established that rapport with their men that comes with a longer service. Where officers had been serving for some time, it seems evident that the men refused to obey them only when threatened directly by the radical elements.

At Pacific Command Headquarters, meanwhile, the staff had been extremely busy. It was not until mid-morning on Sunday that Lt. Gen. Sansom left for Ottawa and Major-General Pearkes could send the last of his senior officers back to their units. He had been kept informed of events in Terrace by Lt. Col. Hendrie, who, to counter the demonstrations staged by the mutineers, asked Pearkes to arrange for a large flight of bombers over the camp, presumably to emphasize that outside assistance could be brought in should the need arise. Pearkes replied early Monday morning:

> Reference your recommendations, I still rely upon loyalty and common sense of men in the Fusiliers du St. Laurent to obey the lawful commands of their officers and to commence making preparations for their move to Valcartier which has been ordered to take place in a few days, rather than to use outside forces which at this time are available to me.[27]

The idea of this display for the benefit of the mutineers was passed on to Ottawa, where it was cancelled. Pearkes, meanwhile, kept Ottawa informed of the reaction to the conscription order in the camps under his command. He asked that several hundred Provost personnel be sent to his command as soon as possible and, as requested, advised his senior officers to arrange to move the battalions in the 6th Division to Eastern Canada prior to being sent overseas. He told Ottawa that Lt. Col. Hendrie and Major Thuot appeared to be handling the situation in Terrace "with tact and firmness", but added, "they cannot take too forcible a line of action as they only have a handful of young officers to back them up." Brigadier A. R. Roy and the battalion commanders were en route from Vancouver to Terrace by train since the weather, always uncertain at best, precluded flying them to the Terrace R.C.A.F. station.[28]

The order to start moving the battalions to Eastern Canada reached Terrace late on Sunday and presented the officers with a ticklish problem. They decided to send an advance party of two officers and fifty men from the Prince Albert Volunteers. The men in this unit, according to the senior officers, were disposed to accept the government's order and although some had joined the earlier demonstration, there seemed to be a smaller core of outright resisters to conscription than in the other units. In addition, the Volunteers' barracks were furthest from the Fusiliers du St. Laurent who, according to rumor, had a variety of weapons to enforce the demands of the mutineers.

On Monday morning the arrangements were complete. Word of the move quickly spread to the rest of the brigade and a large body of men from the other battalions marched to the Volunteers' lines. An officer in the Volunteers described what followed:

> On finding that the advance party had left, they seized three of our vehicles and gave chase. Finding that the train had already left, they abandoned the vehicles on the roadside. Those who did not give chase filled the men's quarters in mobs and pointing weapons in all directions waved the magazines before them to show that they were loaded. Then they told my men in my presence that they would be shot if they attempted to prepare to move.
>
> I then tried to speak to the men and a Bren gun was stuck in my stomach and I was told I would be shot if I did not shut my mouth. On entering the other "C" Company hut, I was thrown out bodily after refusing to leave at rifle point, and warned that any attempt on my part to move the company would mean I would be shot. I then left for the Bn Orderly Room where a number of officers had gathered after receiving the same treatment as myself, and the Orderly Room was surrounded by a mob who were pointing rifles, Brens and Stens at all the windows ordering the officers to leave. We refused to go, and they warned us that they were 3,000 strong and meant business, and that they would shoot if we made any further attempt to move the Battalion or any part of it.[29]

Only two officers and fourteen men in the advance party managed to board the train for Prince Rupert, the two trucks carrying the remainder having been stopped by the mutineers. When it was reported that vehicles filled with armed men planned to cut off the train which was en route to Prince Rupert to remove those who had left the camp, there was some fear that the mutineers might come all the way through. To counter this, the officer commanding the Prince Rupert Defenses gathered together a number of reliable NCOs and officers and sent them to establish a road block at the Galloway Rapids Bridges. As has been seen, the mutineers did not give chase beyond Terrace. What might have occurred at the road block had the two parties encountered each other is interesting to contemplate.

The "picketing" of the Volunteers' lines during Monday afternoon and evening prevented further movement to the trains by the main body of the battalion. Unless the other two infantry battalions could be persuaded to refuse to follow the leaders of the mutiny, nothing further could be done.

Fortunately, on this same day the brigade commander and his senior officers arrived in Terrace. By Monday evening all unit commanders were once more on duty.

In the lines of the 19th Field Ambulance, Captain Belanger had the least number of men to control. His men, or the majority of them, were not sympathetic to the mutiny. During the night of 26-27 November, he planned to have a truckload of ammunition from the ammunition dump in his lines taken away and tossed into the Skeena River, but the mutineers broke into the dump and distributed the ammunition to the men. He did get most of the arms stored there secretly taken over to the R.C.A.F. station. His commanding officer, Lt. Col. P.A. Costin, arrived back in Terrace shortly after noon. Although he and Lt. Col. L'Heureux had been briefed on the situation during the drive from Prince Rupert to Terrace, when he spoke to his officers and trusted NCOs Costin was surprised at the influence the mutineers had achieved in his unit. That afternoon, noting "all kinds of rumours spreading around," he wrote:

> Intimidation have [sic] astonishing effect. Most of our men are afraid to death. No leaders in the Unit. Liaison agents and leaders are supplied from Fus. du S.-L.[30]

At a meeting that evening with Brigadier Roy and the other unit commanders, steps were taken to restore discipline. Until further notice, the men's pay would be stopped and an announcement was made to this effect. For the past several days only the officers, warrant officers and some sergeants had shown up on the parade square at the proper times. Brigadier Roy and his officers believed it was essential to get the men back on parade, no matter how small the group attending might be, nor whether at first they paraded inside a canteen or a hut rather than outside.

Early on Tuesday morning Lt. Col. Costin and Capt. Belanger started to work on their men. Notwithstanding the signs reading "Out of Bounds to All Officers" signed by the "O.C., Zombie Corps", the officers entered the men's barracks and talked to them for some time, persuading them to surrender their arms and ammunition and attend parade. The first break came in Headquarters Company where the men refused to surrender their weapons but agreed to have them locked up under guard in one of the drying rooms. At eleven o'clock about 50 men attended a meeting called by Costin, who stressed that they had nothing to fear from those threatening to shoot. Immediately after noon a special order by Major-General Pearkes was posted in all huts throughout the unit and in all the camps. It ordered all men to obey their officers and NCOs, turn in their ammunition and resume parades and training, and added that sergeants must resume their rank and position. With this was a notice drawing everyone's attention to paragraph 420(c) of the King's Rules and Regulations (Canada), which warned all of the grave consequences to anyone who "causes or conspires with any other persons to cause any mutiny or sedition in any of His Majesty's military... forces..."[31] This was posted and read by officers to all the men. A truck was brought to the door of each hut after the order had been read and, within a short time, all the arms and ammunition were collected and taken away without interference. A mass meeting at the drill hall occupied the attention of the Fusiliers and Highlanders at this time. Later, when some infantrymen came into the lines, all the rifles were at the R.C.A.F. station six miles away and the rest of the ammunition had been dumped into the river.

By the evening of 28 November the "war of nerves", as the war diarist aptly termed it, was almost over. Surrendering the arms and ammunition brought a sense of relief to all ranks. The radio announcement that evening of the Field Ambulance's surrender of their arms "had a most salutory effect", the diarist added, as did the rumor that there was a division in the ranks of the mutineers, "many of who have participated only through intimidation."[32]

Lt. Col. W. J. MacDonald, Commanding Officer of the Prince Edward Island Highlanders, had returned to his unit Monday evening. The order to the men in the battalion to return to normal duty had been posted, but only a minority were willing to go on parade. Nevertheless, reaction was setting in and the day he returned, 21 men volunteered for General Service.[33] Following the plan laid down by Brigadier Roy, Lt. Col. MacDonald called a battalion parade in the drill hall at nine o'clock on Tuesday, but only half the men could be persuaded to attend. It was a chance to talk to them, bring the pertinent section of the Army Act to their attention again, and urge them to return to duties before the situation got worse. He then left to visit the huts. Meanwhile the leaders of the mutiny in the Highlanders, soon supported by those in the Fusiliers, gathered a large body of men and marched on the drill hall to break up the parade there. Major Wright met them and, using all his powers of persuasion, managed to get them to listen to him in the drill hall for almost an hour.[34]

Although unable to get anyone to go on parader that afternoon except the Pipe Band and those who had gone active, Wright had better luck on Wednesday. The difficulty, his company commanders said, "was due to the fact that everyone seemed afraid of everyone else."[35] By calling company parades inside their own huts, however, the officers began to meet with a measure of success, although in some instances abusive language and threats were used against the officers. Then, on Thursday morning, it was learned that the Fusiliers du St. Laurent were returning to normal discipline.

Lt. Col. P. H. L'Heureux, commanding the Fusiliers, met with the men in his battalion for the first time when, on returning from the demonstration in the Volunteers' lines, they went to the drill hall. Here, amidst shouting, he talked to them, listened to their grievances, warned them about the seriousness of their action in both English and French, and directed them to return to their duties. On Tuesday L'Heureux took the same action as the other commanding officers, visiting the men's huts, meeting company "representatives" and ordering his sergeants to replace their stripes. That evening he was able to inform Brigadier Roy that he felt the mutiny was almost over. By 29 November it appeared he was correct. On the morning parade there were a little over one hundred men. Another parade after lunch brought about four hundred men. Except for "B" Company, most of the battalion was on parade by the end of the day and the men were turning in their ammunition. The following day almost all in the Fusiliers and the Highlanders returned to normal duties. By week's end everyone was preparing to leave Terrace according to the orders given five days earlier.[36]

It was not until 29 November that the Prince Albert Volunteers, assured by Brigadier Roy there would be no outside resistance, were able to leave Terrace. Even then the hard core of mutineers within the battalion tried to prevent the departure, threatening to shoot if anyone left the lines and waving loaded rifles and machine guns about as the bulk of the men finished packing. Without outside support, ignored by most of their comrades, and finally driven from the huts where the baggage was being loaded on trucks, these last ditch rebels realized the futility of their protest. As the battalion started to move out, they too gathered up their gear and joined it. The war diarist of the 19th Field Ambulance probably reflected the feelings of most men in the brigade when, on the following day, hearing that the remaining infantry battalions were under control and that the normal routine had returned, wrote: "It all seems like a bad dream."

These events constituted what was probably the largest and longest mutiny ever to occur in the Canadian Army. Its origins were political, not military in nature. Most remarkable about it was that no blood was shed, despite the comparative youth of some of the officers involved. The combination of the Prime Minister's turnabout on conscription for overseas service, the lack of prior warning to prepare the Home Defense men for the new situation, the refusal of Pearkes' request to permit the commanding officers to remain with their units during the critical weekend, and the isolation of the 15th Brigade in Terrace all played their part in creating a situation ripe for protest and disobedience.

Immediately following the end of the mutiny, a Court of Enquiry was held on order of Major-General Pearkes.[37] The findings of the court resulted in demands that prompt disciplinary action be taken. This was easier said than done. Although a few men in the brigade were given prison sentences ranging from one to two years, the need for men overseas and the transfer of units from one military district to another complicated and hampered military justice, as did the very numbers involved. In mid-January, 1945, for example, of 89 men charged in the Fusiliers du St. Laurent, 68 were absent without leave. Later, those who went to the United Kingdom as reinforcements after having taken part in the mutiny were probably thankful when the military authorities there suggested to Ottawa that all charges should be dismissed since the morale and conduct of the former Home Defense men had been good. Morale might suffer, it was suggested, if charges were pressed. They were not.

Notes

1. Directorate of History File 369.009(D3), Department of National Defense, Ottawa (hereafter DND): Report, Sgt. J. Chaharyn, No. 6 Field Security Section to O.C., No. 6 Field Security Section, Prince George, B.C., 12 May 1944. These files hereafter are cited as D. Hist.
2. D. Hist. 112.21009(D.209): DND, "Trends in the thinking of Army Units, Vol. 3, No. 9, September, 1944", prepared by the Research and Information Section, Adjutant-General's Branch.
3. D. Hist., 322.009(D217): DND, Composite Report, Attitudes of Personnel, 13 October 1944, by GSO II Intelligence (Security), Pacific Command, Appendix "B", Comments by O.C., 25th Field Company, R.C.E.
4. The former commander of the Canadian Army, Lt. Gen. A. G. L. McNaughton, replaced the Honorable J. L. Ralston as the Minister of National Defense on 2 November 1944. He hoped, as did the Prime Minister, that his appeal would cause enough conscripts to go active to provide the reinforcements being demanded by the overseas formations.
5. D. Hist., HS322.009(D30): DND, "Abridged Report of the Inspection of the 15th Canadian Infantry Brigade, 13-17 November 1944, by Major-General R. O. Alexander.
6. Exercise "Polar Bear" was a "wet cold" Test held in the winter of 1944-1945. The base of the Force was at Williams Lake.
7. "Abridged Report. . .", op cit.
8. D. Hist., 3221009(D81): DND, "Abridged Report of the Inspection of the Prince Edward Highlanders, 15 November 1944", by Maj. Gen. R. O. Alexander.
9. D. Hist., 322.009(D88): DND, "Abridged Report of the Inspection of the 1st Bn., Prince Albert Volunteers, 13 November 1944", by Maj. Gen. R. O. Alexander.
10. For a more detailed account of the conscription crisis as it affected Pacific Command, see R. H. Roy, "Major-General G. R. Pearkes and the Conscription Crisis in British Columbia, 1944", in British Columbia Studies, No. 28, Winter, 1975-76, pp. 53-72.
11. D. Hist., DND, File HOS 20-6, Vol. 81, Telegram, Pearkes to Major General A.E. Walford, 23 November, 1944.

12. D. Hist., DND, File PCA-787, Telegram, Top Secret, G.O.C.-in-C. Pacific Command to Div. and Formation Commanders, 1547 hrs., 23 November 1944.
13. Public Archives of Canada, DND Records, File 3545, Vol. 5, (R.G. 24, Vol. 2654), Court of Inquiry, 3 December 1944 (Henceforth quoted as Court of Inquiry); Statement by A/Capt. G. A. Anthony.
14. Court of Inquiry, Evidence of Major H. J. Thuot. The brigade major, Major R. L. Williams, added that as it was near the end of the month, the quota of beer for all canteen had been nearly consumed. Pay day was a week away.
15. Court of Inquiry, Letter, Major R. W. MacMillan to H.Q., 15th Cdn. Inf. Bde., 26 November 1944. Pacific Command did its best to keep formation commanders informed about the actual state of affairs. The Chief Censor of Publications was sent from Ottawa to Vancouver to assist the local representative. During the next few days there were several occasions when false or misleading radio reports had a direct effect on affairs in Terrace. (D. Hist., DND, Adjutant-General's File BDF 45-34, Letter, Wilfred Eggleston to the Minister of National Defense, 7 December 1944).
16. Court of Inquiry, Appendix "I", Report of Captain C. H. Belanger, 19th Cdn. Fd. Amb., R.C.A.M.C.
17. Thuot, *op. cit.*
18. Court of Inquiry, Evidence of Lt. Col. W. B. Hendrie.
19. Hendrie, *op. cit.*
20. MacMillan, Letter, *op. cit.* It was estimated that about 50-60 Volunteers took part in the parade.
21. Belanger, *op. cit.*
22. PAC, RG 24, Vol. 2654, DND Records, File 3545, Vol. 5, Report on Terrace, B.C., D.O.C., M.D. 5 to the Secretary, DND, Report by Lt. Col. P. H. L'Hereux, Appendix "B", Men's Demands.
23. There were only four sergeants in the battalion who were "GS". At this time there was only one sergeant major. He was "active" also. The remaining sergeants and warrant officers, i.e. about 75-80%, were Home Defense.
24. Court of Inquiry, Evidence by Major H. J. Thuot.
25. Thuot, *op. cit.* It is interesting to note that a few sergeants refused to take down their stripes but decided then and there to volunteer for active service!
26. Court of Inquiry, Evidence by Major J. S. Wright, Acting Commanding Officer, Prince Edward Island Highlanders.
27. Telegram, Pearkes to Hendrie, 27 November 1944, Inquiry, Appendix. The phrasing of the telegram is such that it appears both officers were unsure whether the message might be read by the leaders of the mutiny.
28. D. Hist., File HQS 20-6, Vol. 81, Telegram, Pearkes to Defensor, 26 November 1944, DND.
29. D. Hist., File 746.009 (D8): DND, "Prince Albert Volunteers, Reports on Disturbances", Statement by Acting Captain G. A. Anthony, 20 December 1944.
30. Court of Enquiry, Exhibit "J", "Report from Lt. Col. P. A. Costin, 3 December 1944. Among the rumours was a report that the mutineers intended blowing up the Sergeant's Mess and the Skeena Bridge. (W.D., No. 19 Canadian Field Ambulance, RCAMC, November, 1944). One officer reported the mutineers had removed explosives from the magazine but later replaced it.
31. This has already been posted in unit Daily Orders on 25 November 1944.
32. W.D., 19th Canadian Field Ambulance, R.C.A.M.C., 28 November 1944.
33. W.D., The Prince Edward Island Highlanders, 27 November 1944.

34. Court of Enquiry, *op. cit.*, Evidence of Major J. S. Wright. Wright reassumed command of the battalion that evening when Lt. Col. MacDonald left for Prince George.
35. *Ibid.*
36. W.D., Fusiliers du St. Laurent, 27-30 November, 1944. L'Heureux had told his men on Tuesday that unless they were back on parade the following day, a flight of R.C.A.F. aircraft would fly over the camp. The airplanes, he said, carried no bombs, and anyone seen outside with a weapon who refused to lay the weapon down would be arrested as a mutineer. (Court of Enquiry, *op. cit.*, Evidence of Major H. J. Thuot.) The unit war diarist claimed that poor weather prevented the flight of the aircraft on 29 November. No mention is made of the possibility of such a demonstration by any other unit war diarist. Pearkes flew to Terrace on the 29th, however, returning to Prince George the same afternoon after conferring with Brigadier Roy and others.
37. Pearkes had flown up to divisional headquarters late on 28 November but was dissuaded from continuing on to Terrace. He remained in Prince George for two days.

TECHNOLOGY AND TACTICS

Sans Doctrine: British Army Tactics in the First World War

Dominick Graham
University of New Brunswick

It seems to be generally agreed that senior First World War military leaders were incompetent. Prima facie evidence of their incompetence lies in the huge casualties resulting from their blunders and in their ignoring the predictions of civilian and military writers before the war that a future conflict would be a bloody stalemate.[1] Any catalogue of their major errors would include the faults in conception and execution of the Schlieffen Plan, the Grandmaison doctrine of the *offensive á outrance*, and the administrative muddles of the Russian General Staff. Haig's obstinate perseverance with the Somme and Ypres campaigns long after they could succeed is a focus for British criticism.

The reputation of British leadership has rested on the broad shoulders of the Army's enigmatic chief because popular writers, having found him to be a specific target which they could not miss, continued to attack him as a symbol of military leadership. Their attempts to infiltrate beyond that Verdun and to widen their attack have generally broken down in a morass of subjectivity, contradictions and myth into which they have led the public. They have praised the British Expeditionary Force of 1914 as a notable exception to a general rule that British expeditionary forces are disastrous, but they have condemned the professional soldiers who built it.[2] They have depicted citizen-soldiers as naturally imaginative, progressive and democratic, epitomized by the Australians, New Zealanders and Canadians whose divisions usually appear to have been successful.[3] If Britain's own citizens were less successful, it was because their leaders were unable to adapt to a form of warfare that they had not anticipated and remained tied to the obsolescent attitudes of their class. On the other hand, the German Great General Staff was the epitome of professionalism where the British staff was amateurish; unimpeded by its extreme class attitudes, it was more successful in the field. The argument for German superiority seems to hang on comparisons between the casualty figures of the two armies and between the amount of territory gained by the British in 1916 and 1917 and by the Germans in Russia and, in 1918, in France and Belgium. These criteria are seldom challenged or even discussed intelligently in the light of policy, developing tactical doctrine and practice, and differing field conditions. In tactical innovation the Germans also seem to have been awarded the palm. The tank is not permitted to be the exception that proves the rule because its creation was largely due to amateurs while professionals bungled its employment.[4]

Professional historians, as opposed to popular writers, tend to temper criticism of British commanders and staffs with the observation that the post-Boer War reforms came too late to provide enough leaders for a huge citizen army. The single peacetime corps and the belatedly reformed staff college were inadequate to staff the score of corps and sixty divisions eventually deployed in France and Belgium alone, and training for high command could not be given in the small peacetime army. Furthermore, the parsimony of the Treasury, overseas commitments, and the retrenchment and reform policies of successive governments handicapped the General Staff in preparing the Expeditionary Force for any continental war, let alone the positional war of materiel that it had to fight from the winter of 1914 onwards. Yet, granted these extenuating circumstances, the leaders are usually blamed for failing to grasp the nature of positional warfare and the importance of its dominating weapons, the machine gun and heavy artillery. Had they done so, it is implied, they would surely have contrived tactics and weapons to defeat the German Army sooner and with fewer casualties. Nor would they have entered the war believing in that nostrum, the offensive spirit. Brian Bond's mild yet damning verdict on the British commanders of 1914 is that they were late-Victorians confronted by a twentieth century war.[5]

Presumably Bond means that collectively the decision-makers of 1914 did not grasp the effects of modern fire-power on flesh and blood. That conclusion prompts a counter-comment that opens the door to wider issues. The British cabinet of 1914 had neglected to study modern war or the requirements of the land forces to be committed to it. Subsequently, faced with the effects of pre-war policy, the soldiers were unwilling to shoulder its compulsions alone. On the other hand, David Lloyd George admitted no responsibility for the casualty lists or for British unpreparedness. The disputes between Frocks and Brass Hats were created by incomprehension of each others' pre-war unadmitted sins of omission and commission. After 1918 the legacy of distrust between soldiers and politicians was passed to those who were thrust into the Depression and then, even more unprepared than in 1914, into the disasters of 1940 and the years of retreat and humiliation that followed. Then it seemed to the soldiers, professional and citizen, that the sacrifices made in 1914-18 had been thrown away by incompetent politicians during the peace. Not only had they to fight another war to end war but one to rid the country of social and economic injustice.[6]

It is no wonder that a school of disenchantment[7] dominated opinion about the First World War for so long, resisted objective analyses of the fighting, and declared a plague on both houses and on war in general. By the 1960s, however, it was obvious that the general malaise pervading British society could not be attributed to the First World War and its cruel aftermath alone. It was also the expression of more widespread dismay over an apparent paradox: after two victorious world wars and twenty-five years of reforms, Britain was apparently weaker and her internal problems exacerbated. Historians have responded by examining the history of the two wars as episodes in a national decline that began before 1914. Military historians have begun to compare the two wars instead of contrasting them. But they are only in the ante-room of wisdom, for few have resisted the tendency to use hindsight from the Second War to judge the First, or to use the methods of the First as a pejorative standard in judging some operations in the Second. Their judgments of the commanders of the two wars are not yet supported by rigorous examinations of their tactics and of the policy behind the tactics. Indeed, our knowledge of tactics in the two wars as a developing study, with social, political, technological and ideological ingredients, is spotty and clouded by emotion. The emotion may do us credit but it has contributed to prejudices, e.g. that the First War was not only bloodier but also less progressive than the Second and that its commanders were of a lower calibre.[8] That may or may not be so; but comparisons and contrasts are worthless unless the accounts from which they are drawn are written with common terms of reference and embrace the periods of peace preceding the wars. We are still waiting for such accounts.[9]

This essay is an outline sketch of material that may be included in assessing the first half of the period. It is written in a form and with themes that may be continued into the peace and through the Second War, giving unity to the military history of a period that has lacked any to date.

The general limits that policy put on the readiness of the British Army for continental war are well known. The Expeditionary Force, consisting of only six infantry divisions and one large cavalry division, was required not only for a continental war, to which it might never be committed, but also for its primary role as imperial reserve. In addition, its units continuously trained the men and found the drafts to keep the overseas army at peace establishment. In fulfilling this imperial commitment, at a time when the army as a whole was not at full establishment, the units of the Expeditionary Force could never take the field for manoeuvres in the UK fully trained and to strength. On mobilization for war, regular battalions received two reservists for every man serving with the colours. Some reservists would have served only three years with the colours, many would be unfamiliar with the new organization and weapons, and all would be unfit for marching.[10] The Army Council was caught in a vicious circle due to conflicts between the continental and imperial commitments. The peacetime demands of the overseas army, by draining home units of trained men, increased the reserve increment that they required on mobilization for continental war. To ensure that the Expeditionary Force would be efficient on mobilization, the Army Council reduced the percentage of reservists in it by allowing men with the colours to extend their service in peacetime. This measure was made more essential by falling enlistments and an increase in the number passing to the reserve in the years preceding 1914, but it conflicted with the principle of building a reserve adequate for continental commitments.[11] Nor could the Army Council depend on the Territorial Force as an immediate reserve for continental war. Embodied for home service only, six months would elapse before its fourteen divisions would be ready to face a European enemy, for there was not enough money in peacetime to provide it with adequate training centers, cadres and equipment without depriving the Expeditionary Force of essential men and weapons. In consequence, the Territorial Force had obsolescent weapons, was below strength and on the old battalion organization, and had lax training standards, although its men were potentially of high military caliber.[12]

The political facts, above all others, that determined the conditions under which the Expeditionary Force was eventually committed in 1914 were that the Haldane reforms were accepted only if they saved money, and that the Cabinet, in peace, did not commit itself to send troops to the Continent. The first fact led to annual reductions in the army budget from 1907 until expenditure on the new air service raised it again for the first time in 1913.[13] The Army Council felt compelled to offer loss-leaders to the Treasury for every reform and innovation that it introduced during these years.[14] The second fact restricted joint planning with the French to arrangements for transporting the Expeditionary Force from the ports to its concentration area. Furthermore, it made it inadvisable for the Cabinet to discuss the probable course and duration of land hostilities. For if it examined the implications of even the limited and qualified commitment envisaged, ministers' hands might have been tied by those opposed to any commitment at all. As involvement on the Continent would have been unpopular, Asquith was well advised to leave the Army in the policy vacuum in which it rested until August 4th, 1914.

Although the Government's Continental policy was ambiguous, the Army focused its attention, as far as possible, on that commitment from the Russo-Japanese War onwards. Continental tactics were dominated by French and German theories; the British sought a suitable middle way between them.[15] In the last analysis, by 1912, the tactical problem was how to cross the last four hundred yards of fire-swept ground to close with the enemy without suffering crippling losses. The French believed that since neither artillery nor infantry weapons could decisively weaken a well-entrenched enemy, numbers, speed and elan were the keys to success in the final assault. Therefore they were prepared to accept heavy casualties. The Germans, on the other hand, believed that preparatory artillery concentrations could be so effective that frontal infantry assaults might be relatively inexpensive. But they did not believe in repeating the assault if it failed, preferring flank envelopment. Naturally all armies emphasized the need for high morale in units that were required to advance under heavy fire.

Some observations should be made on the debate about the assault. First, the success of the French 75 mm field gun, introduced as early as 1897, had largely shaped French infantry tactics. Neutralizing shoots from open positions suited its high muzzle velocity and rate of fire. It was at its most effective at short range against an enemy in the open or lightly entrenched. On the other hand, it was ill-adapted to indirect shooting or the destruction of entrenchments, where deliberate methods and heavier calibres were required. As the French had no howitzers for that purpose in the divisional artilleries and were short of them elsewhere, they advocated the 75 mm *rafale* to demoralize their enemy before attacking. The Germans, however, had developed howitzers, using field, medium and heavy calibers to support their divisions. They advocated using them deliberately to destroy entrenched infantry. The value of good entrenchments was not questioned by either army. Nor was there any doubt that attacking unshaken infantry in them would be costly. What was in dispute in England was the ability of the continental or their own artillery to kill or wound the occupants immediately before they were assaulted. The French set out to neutralize, the Germans to destroy them. Who was right? The British infantry preferred French methods, since in practice they promised closer support and because they distrusted indirect shooting.[16]

The machine gun was at the center of the other important aspect of the debate about the assault. The Manchurian War had restored its reputation after a poor showing in South Africa. In the years immediately before 1914, the Germans started to increase the establishment of machine guns in the army from one six-gun company in each infantry regiment of three battalions to a company of six guns in each battalion. The 650 battalions serving with the colors were in the midst of this change in 1914.[17] The course of the machine gun debate in the British Army illuminates British ideas on the assault and on tactics in general.

Trials at the Musketry School at Hythe revealed that at 600 yards two Maxim machine guns could annihilate a battalion advancing in the open in extended order in one minute, provided the men did not go to ground. The School defined annihilation as 60% casualties. It made two direct deductions from this trial and others that tested the effectiveness of aimed rifle fire against such an attack. First, the battalion did not require more than two machine guns in defense; secondly, and conversely, the manpower of the battalion ought to be reemployed to achieve greater firepower in the attack. As the Maxim was a cumbersome weapon, unsuited to the assault, it suggested that automatic rifles or some other form of light automatic should replace some "bayonets".[18] Mortars and hand grenades were considered as an addition to the battalion firepower, but in their rudimentary state they found little favor and money was lacking to develop them.[19] Several automatic rifles and light automatics were presented but all failed the trials which were held when the Lee Enfield MKIII rifle, resighted for Mark VII ammunition, was about to be adopted.

Introducing a new rifle seemed a sufficient undertaking for the Army and the Treasury, particularly as the School of Musketry's ideas introduced a principle which found little favor with the infantry: that mass fire and support weapons should be increased at the expense of the rifle. The infantry had made the rifle a weapon of great precision since the inaccuracy of rifle fire in Manchuria had revealed the potential advantage of straighter shooting.[20] Moreover, all its manpower was needed in the assault and it was feared that the bayonet strength of the battalions would be whittled away if specialists were introduced whose value in the assault had yet to be proved. Thus the only positive result of the School's initiative was that, belatedly, the lighter Vickers machine gun replaced the Maxim in the cavalry. It was agreed that when the Vickers had been built in sufficient numbers the infantry would have it too.

Although the infantry was directly concerned in the debate over artillery fire support, its only contribution was pressing the Gunners to use French methods to neutralize or destroy enemy machine guns and stigmatizing indirect shooting as "slow". Belatedly the Gunners began to teach the infantry the advantages of the latter, but in the field they could not undertake to destroy an entrenched enemy, let alone individual machine guns, because only the eighteen 4.5 inch howitzers and the four 60 pounders were provided with high explosive shell, and they had no means of directing them as a single fire unit. The opinion in the field artillery, soon after the 4.5 inch howitzer was issued in 1910, was that its scale should be increased and that the 18 pounder field guns should have a high explosive shell too. Nothing came of the former proposal, and the fuse for the 18 pounder shell ran into difficulty. The artillery command organization was notoriously too small, and its communications too slender to concentrate divisional fire or control it, let alone that of the tiny army artillery. The largest divisional artillery fire support unit was the battery. But even battery commanders lacked the telephones and wire to link their guns, if they were in concealed positions, to their observation posts and the infantry whom they were supporting. Consequently techniques for supporting the infantry were only suited to small engagements in open warfare.[21]

The course of the debate over artillery support and over infantry automatic and support weapons reflects the principal weakness in the British Army of 1914. The big divisions, introduced only in 1907, had not yet fully developed into teams of all arms.[22] Their units still thought as separate families, each with a history, proprietary functions and weapons. The infantry battalion at home still looked to its overseas battalion, its depot and its regimental colonel, and to the rifle as the foci of its interest; the field and horse artillery batteries remained the elite historical and tactical units of the Gunners at the expense of the recently created divisional commanders' Royal Artillery, the field artillery brigade commanders, and the Royal Garrison Artillery batteries of heavy guns and siege howitzers. The infantry and artillery schools that are responsible for developing tactical doctrine in a modern army did not exist in their present form. Divisional commanders were at liberty to interpret the cliches of the manuals as they wished, and few agreed. More to the point, none rose above the arm in which they had been trained. Yet the age-old tactical problem of closing with the enemy had an equally venerable solution; close cooperation between divisional weapons and more of them. The doctrines of the arms had to mesh; it was not enough that each throw its contribution on the butcher's scales, like assorted hunks of meat. In 1914 the General Staff had not imposed one common tactical doctrine on the Expeditionary Force. Only arms doctrines existed, like the trust in aimed rifle fire. By 1915, few marksmen survived in battalions, as the School of Musketry had predicted, so the lack of support weapons would be sorely felt. The neglect of Corps and Army artillery, and of the techniques for managing its reserve firepower, would bring its revenge too.[23]

In short, the Expeditionary Force was ill-prepared to fight divisional battles, let alone those requiring the integration of divisions and corps. Perhaps the best explanation for its condition lies more in the politics and sociology of organizations than in a lack of sound military ideas. A critical mass of middle-piece, staff-trained officers had not yet been created to ensure the adoption of methods that were widely discussed and advocated.

The battles from Neuve Chapelle in 1915 to Cambrai in November 1917 were episodes in the painfully slow building of an army envisaged by many progressive soldiers in peacetime but which money, policy, and time had put beyond their reach. Time, after all, had been lacking. The new General Staff officers required time to acquire power with seniority and to educate their juniors. Once the war started, time was still the commodity they would need most but be denied, and promotion would come all too quickly to the few of them who survived the early crises.

When the Expeditionary Force was committed to the Continent its commanders expected to play a subordinate role in operations that had been conceived by their ally before the war without their knowledge or agreement. While co-operating with the French, they had to preserve the regular divisions from destruction, subsisting on the regular army reserves for six months until the Territorial divisions were ready. But so heavy were Sir John French's casualties that by December 1914 his army was hanging on the ropes, although it had been reinforced by an Indian Corps, some Territorial battalions sent ahead of their divisions, and new divisions formed from regular battalions recalled from overseas. In the circumstances he was content to dig in, rest, train and wait for reinforcements. The Germans, too, were ready to entrench themselves, after unsuccessfully committing reserve formations judged unfit for difficult operations, while they turned their attention to the East. The positional warfare that continued, to a degree, from then until the battle of Cambrai in 1917 was initiated less by machine guns and heavy artillery than by the will and need of the exhausted combatants. The weapons, men and wire that were accumulated in each succeeding month thereafter simply consolidated the form of war to which both had turned as a temporary resort.[24]

It was generally agreed by the British commanders at the beginning of 1915 that they would have neither the trained men nor the weapons to play a role in breaking the deadlock that year. In the winter, when morale in the trenches was low, the corps staffs recommended that only tactical operations, retreats as well as advances, should be undertaken in the coming year in order to gain experience and establish a line from which the new armies might take the offensive in 1916. 1915 was to be a year for training and the accumulation of materiel and men. But this strategy was unacceptable to the French, who were determined to rid their country of the invader before his manpower advantage became decisive. The British and Russian armies were therefore committed to a series of operations in 1915 for which neither was ready. It was not until 1917, when the battles of Messines and 3rd Ypres were fought, that the location and timing of operations were selected by the British command. Nor had the Army, until then, acquired enough weapons and munitions of sufficient quality, and techniques for their employment, to take the initiative and to escape from the tactics as well as the strategy of attrition.

The battles of 1915 and 1916, when French direction prevailed, built up the experience of the Corps and Army staffs in managing positional warfare. They played the major roles in the break-in battles of those years. As the divisions that they controlled were semi-trained they kept them on a tight rein and conducted the fighting according to bureaucratic rules that left no room for initiative and were also self-perpetuating. For the battalions, and, to a lesser degree, the observers and battery commanders of the artillery, suffered such casualties that experienced men did not accumulate to train units to exploit the semi-open conditions which often existed temporarily after a break-in. Indeed, the laborious techniques of artillery support by timetable, once mastered, became ends in themselves and determined infantry tactics. Furthermore, the regular army itself had had to learn the specialized techniques of trench operations before teaching them to the new armies which were originally "trained" for open warfare. The conditions upon which the army entered the war ensured that neither the men nor the time existed to retrain a whole army to fight in the open while the deadly business of the break-in had still to be mastered.[25]

The years 1915 and 1916 were well spent in overcoming some of the material handicaps under which the Expeditionary Force entered the war but the accompanying doctrine and techniques were less easily acquired. The infantry weapons that the School of Musketry had proposed were introduced and their numbers were progressively increased. Yet the doctrine, techniques and specific ordinance for hitherto unfamiliar weapons had to be developed. Even in 1918, German *Stosstruppen* used mortars more effectively than British infantry, while British artillery mortars were still too immobile. Lewis guns, eventually issued on a scale of one in each platoon, were invaluable for holding objectives against counterattacks and in defense, but they were not often used imaginatively in the attack. The battalion bombers were the elite of the trenches, but they suffered cruelly until they had enough bombs of modern design. In the semi-open conditions of the second half of 1916 the short range of the hand-grenade made it less useful and the infantry had to be taught to use their neglected rifles.[26]

Not until the end of 1917 had the artillery developed most of the techniques that were still in use in 1942. However, in 1916 on the Somme, although the number of guns and the amount of ammunition were sufficient, there were still deficiencies in the quality of both. As many as 30% of the shells may not have exploded and 25% of the guns were generally out of action awaiting repair. The accuracy of the guns also suffered from faulty fire direction techniques. Shooting to the map was inaccurate because survey was still imperfect, meteor telegrams were not received frequently enough, calibration was largely guesswork, propellant lots were mixed, and gun wear became extreme as battles progressed. Observed shooting was often impossible because of broken communications. Counter-battery fire, which depended on all these and on continuous air observation as well, was seldom accurate; thus enemy shellfire caused enormous casualties.

On the first day of a battle, when the artillery had been given time to locate the enemy's infantry and guns and could use weight and volume of fire as a substitute for precision, it gave a good account of itself. Its defects were fatal when the battle moved beyond the forward defenses. For it could not give quick, impromptu fire support, nor rapidly fix new hostile batteries. Tactical surprise was impossible to exploit when rigid, predictable and relatively dependable fire plans had to be preferred to those that, hastily arranged, were undependable.[27]

Tanks arrived on the battlefield in September 1916, when the artillery was still not able to play its part in achieving surprise and the infantry was still not combining its new weapons effectively in semi-open fighting. Potentially, the tanks would provide the impromptu fire support that the infantry lacked and the artillery could not provide. Actually, the vehicles had not the mechanical ability to traverse ground churned into a quagmire by bombardment, their crews lacked tactical experience, and the other arms had not mastered their own crafts sufficiently to cooperate with them.

The Mark IV tanks and their successors performed better in 1917, and there was an improvement in the infantry tactics of a few divisions.[28] More important, by the end of 1917 the artillery had overcome many of its shortcomings. At Cambrai the artillery used gas shells in large quantities and the new 106 instantaneous fuse, and predicted shooting, in conjunction with tanks, by making registration of the targets unnecessary made a preliminary bombardment inessential. The tanks and infantry were able to achieve surprise and to exploit it, marking the first time since Neuve Chapelle, in March 1915, that surprise was the principle of a battle. In the meantime the Germans, using improved artillery techniques and better infantry tactics in the absence of tanks, had returned to the same idea.

The British took their first lesson in the science of positional warfare, and their first step towards breaking out of the deadlock, at Neuve Chapelle in March, 1915. The attack, on a two corps front, was a tactical surprise but failed to reach its objective, the Aubers Range. Only the village of Neuve Chapelle, which had formed a German salient, remained in British hands. The operation not only introduced new techniques, such as briefing by air photographs and wire cutting by single guns at short range, but it was the source of two schools of thought about future battles.

Sir Douglas Haig, commanding First Army, had given the main role in the battle to Sir Henry Rawlinson's IV Corps. Haig wished surprise to be exploited to the hilt. He therefore settled a difference of opinion among the various Gunners about the duration of the preparatory bombardment in favor of one lasting only forty minutes. The assaulting 8th Division and IV Corps staffs doubted whether the artillery concentration could achieve its purpose in so short a time. Haig's plan was that the assault brigades should penetrate straight to the third German defense zone on the ridge, leaving reserve brigades and the flank divisions to clear up behind them. Rawlinson, on the other hand, wished to pause on the Neuve Chapelle line, using it as a bridgehead into which he would move some guns to support the second phase of the attack by the reserve brigade. He was overruled, but retained the capture of Neuve Chapelle as the first phase, the objective for the two leading battalions of the right assaulting brigade through which he leapfrogged the rear pair to continue the advance. There would be a slight delay while this manoeuvre was accomplished.

The initial setback was that the artillery failed to cut the wire in front of the left assaulting brigade of 8th Division or to damage the trenches sufficiently. That brigade was held up and suffered heavily. The guns that were responsible were reported to have arrived in position too late to carry out registration. After a delay in Neuve Chapelle, the second pair of battalions of the right brigade debouched from the village to be caught in both flanks by machine guns, not only because the brigade on their left was held up but because the Indian Corps on their right had not advanced very far. Haig attributed the failure to slowness in committing the 8th Division reserve brigade and lack of initiative among battalion and company commanders. They were too anxious about their flanks and obsessed with linear tactics when they ought to have been penetrating and enveloping successive enemy positions to help each other forward. In IV Corps, where Rawlinson was caught on the wrong foot after the battle by having to admit that he had blamed the 8th Division Commander, General Francis Davies, instead of his own slowness in committing the reserve brigade, there were other opinions about what had gone wrong. The insufficiency of the preparatory bombardment was blamed, but rather than question the feasibility of Haig's concept of a deep penetration to exploit surprise after a short bombardment, lack of ammunition was made the scapegoat. Yet it was observed that the guns had failed to strike targets of opportunity after the set-piece bombardment ended because communications had broken down. An attempt to use pack artillery for close support had foundered because of mud and blocked "roads", and the 8th Division mortar troop, a new venture, had very soon exhausted its cumbersome ammunition and its men. Since impromptu fire support was impracticable, Rawlinson concluded that he would conduct future attacks in phases with a prearranged fire plan for each and that surprise might be attained by heavier artillery concentrations.[29]

After Neuve Chapelle, the term "defensive flank" expressed a prevailing dogma that attacks on narrow fronts were doomed to wither under artillery and machine gun fire from the flanks. At Loos, in September 1915, the frontage of the attack was widened to establish flank-guard positions, although it increased the number of troops employed initially and reduced to zero those available to exploit success. Moreover, despite the work of the engineers and pioneers at Neuve Chapelle, reinforcements had been delayed by traffic congestion. A wider front gave the commander the benefit of a deeper bridgehead with better communications, but the step from local attacks for tactical objectives to larger operations to threaten the enemy's strategic communications was not accompanied by a development in grand tactics. Although the expression "the break-in" was not used until after the war, it was considered to be the first phase of a two phase operation. But it included the period of attritional fighting that would precede the second phase, which was the breakthrough by the cavalry. The nature of the attritional fighting, its duration and particularly its status as a separate phase that required different methods from the break-in, remained in dispute not only between Haig and Rawlinson but between one division and another. It seems that Rawlinson was convinced of the need to fight the break-in on a broad front according to his own methodical format, but that he had not perceived, by the summer of 1916, that a separate phase of "dog-fighting", using different methods from those of the break-in, was both needed and possible before a breakthrough could be achieved.[30]

When Rawlinson, as Fourth Army Commander, undertook the battle of the Somme, he was still the exponent of the deliberate battle in two phases, and he was still at odds with Haig. By then Haig had been converted to the necessity for a longer preparatory bombardment, although not to the same extent as Rawlinson. Perhaps his new artillery adviser from Fourth Army, Major General Noel Birch, had been responsible for that and for his conversion to the belief that counter-battery fire was as important as bombarding the enemy infantry. But he still insisted that phased repetitions of the original break-in were, in aggregate, more expensive than exploiting the original shock by an immediate deep penetration. As his plan envisaged the seizure of tactical areas rather than continuous German defence lines, he disagreed with Rawlinson's uniform deployment of strength along the flanks of the German salient between the Somme and the Ancre and his methodical reduction of the three German defense zones. He preferred that Rawlinson select *Schwerpunkts* opposite the vital areas and place General Hubert Gough's reserve behind them. Rawlinson argued that there would be time to exploit a favorable thrust line when it appeared, but that he expected to be held in front of the German second zone (except perhaps on the left, where it was close to the front) and anticipated mounting another set-piece battle before advancing to the third zone for a final battle for the breakthrough. The Fourth Army staff knew that the sophistication of German entrenchments declined from front to rear and that breaking them would be progressively easier if they could mount fresh attacks before the Germans had time to dig and wire.[31]

The Somme battles showed that although Haig was generally wrong in expecting deep penetrations on 1st July, Rawlinson's plan was too rigid to allow him to exploit the success of his XIII and XV Corps on the right. Later, Haig's conception proved unsound in that, despite overrunning embryo trench systems in the quasi-breakthroughs of July 14th and September 15th, his divisions were unable to manage the transition to open warfare, or even to conduct semi-open warfare successfully, due to lack of training and suitable tactical doctrine. Rawlinson was more aware than Haig of the effect that the appalling physical conditions were having on the battles. The chaos of almost impassable, heavily shelled tracks through acres of lip to lip shell holes slowed reinforcements, relief and supplies, prevented the artillery from following successes, and made the infantry arrive late, exhausted and depleted at its start line. Then it was hard for them to locate their enemy in the wilderness. There the defender had an advantage, for the British usually occupied positions marked on German artillery maps, whereas the discontinuous refuges of the German infantry were hard to locate on the ground, even from air photographs. Minor operations to straighten the line and to protect the flanks of later set-piece attacks behind the new creeping barrage often proved more expensive than the major battle. Rawlinson's experience confirmed his opinion that progress lay in perfecting the set-piece battle at all levels so that the casualty account was consistently in the black.[32]

The Somme exposed many artillery weaknesses. Efficient counter-battery work, in particular, depended on the Gunners correcting the technical problems. Yet, because infantry soldiers were inclined to underrate counter-battery fire that they could not see, despite suffering casualties from German shells, corps staffs varied in the attention they paid it. The same was true of the training of infantry section and platoon commanders for semi-open fighting. In some divisions, for instance Maxse's 18th Division and in his XVIII Corps later, the schools emphasized small unit teams using the rifle (sadly neglected in the trenches), the platoon Lewis guns, and cooperation with tanks, mortars and artillery, But, as before the war, the lack of central tactical schools deprived the Expeditionary Force of leadership in the search for tactical solutions. Haig maintained the principle that general officers commanding were responsible for training along the lines laid down in General Staff manuals and periodical literature; he did not appoint a Director General of Training to guide the myriad schools at all levels in the Expeditionary Force until July, 1918. Long before that, the movement away from continuous trench systems, which had started after the first two weeks on the Somme, had gone so far that the principles of surprise and efficient small unit tactics had become keys to success recognized by the better divisions.

Messines, in July 1917, marked the high point in the tactics of limited advances by set-piece battles, and a turning point in Western front tactics. Artillery fire on the Somme had forced the Germans to occupy discontinuous defences of shellholes, reinforced ruined villages, redoubts in woods and concrete pillboxes. In 1917 the huge artillery concentrations on forward areas and the increasing effectiveness of British counter-battery work, already demonstrated at Vimy in April, compelled the Germans, where the ground would allow, to organize their defenses in at least three zones arranged to a depth of about seven miles. The British Artillery was to be forced to redeploy to defend the ground won in the first phase of an attack, for German counterattack forces would be located out of range of the preparatory bombardment but close enough to strike the attackers as they arrived, disorganized and weakened, at the back of the first zone of defense. That, at least, was the theory. But at Messines, the terrain compelled the Germans to hold the first zone more strongly than they desired. They lost so heavily from the mines and the bombardment before and during the battle that the British were able to walk through to the back of their forward zone and reorganize in time to pulverize their counterattack. The Germans learned that against such well organized, limited attacks they must hold the front zone with fewer men and more machine guns and with field guns dispersed in open battle positions hidden from the air and not targeted so easily by the British counter-battery organization. Counterattack forces would be held further forward so that the British had less time to organize on the objective. This method was tried in the earlier battles of 3rd Ypres, but British barrage fire swept back and forth in considerable depth, as well as creeping forward with the attacking infantry, finally becoming a defensive curtain in front of the objective through which German counterattackers had to pass. Trial and error convinced the Germans that the forward zone might be lost in a set-piece attack and that they must rely on a network of machine guns and mortars, often placed in concrete structures, and on deliberate counterattacks to inflict unacceptable casualties on the British.

The limited attack was made into a science by General Plumer at Messines. The balance of casualties would have been still more in favor of the British had the battalions that overran the Wytschaete-Messines ridge not dug in on the skyline and forward slope around the old German positions, where they were heavily bombarded. Haig's plan was to continue the same limited methods in August at Ypres, where the Germans could not afford to retire, and seize the opportunity, which he was convinced would be offered to him, when the Germans broke under the hammering. A factor in Haig's subsequent undoing was that Fifth Army's artillery planning was not as enlightened as in the First, Second and Third Armies and that its guns were overlooked to an extent that put them at a grave disadvantage. Nor had Gough air superiority, so his congested gun areas were continually bombarded by air observers. Therefore, the most important advantage that Plumer's infantry had enjoyed was denied to Gough's, for the gunners fought only a drawn battle with their opposite numbers. Furthermore, bombardments, which had made the Somme battlefields a morass and strained logistics there, turned the Ypres battlefield, during the wettest August for eighty years, into lakes of liquid mud. Mud, water and German artillery, rather than ground opposition, strong though it was, destroyed the August timetable. Fine weather in September was followed by rain again in October and November, when Haig finally ran out of time. Much has been written of Haig's error in continuing the Ypres battles beyond the first week of October. Should he have selected the salient for his greatest offensive? A more potent criticism is that he ought to have used the time to train his divisions to exploit the surprise that the tanks and new artillery techniques had made possible. But it was impossible to train for the new warfare while the old-style battles continued, because they consumed divisions and their rigid artillery methods put the infantry and the tanks in a tactical strait-jacket, preventing the development of new methods.[34]

If his diary is any indication, Haig was not aware of the possibilities of a surprise attack by massed tanks in conjunction with an H-Hour bombardment of communications centers, headquarters, gun positions and forward defences without any preliminary registration or bombardment. Nor was Maxse's campaign to improve small team tactics in the infantry mentioned in its pages. After General Byng presented his plan for Cambrai, Haig noted the operation as a breakthrough by cavalry with a force of all arms relieving them at the objective. He gives no inkling of understanding that the fighting had returned to a point where surprise was again possible, or that this was the significance of the battles of Riga and Caporetto (as it would be at Cambrai). Rather than a blueprint for future successes, Haig saw the battle as a byproduct of 3rd Ypres, using methods which had no depth of support in the army as a whole.[35]

Most of the divisions at Cambrai had been mauled in the autumn fighting in the Salient before undergoing up to a month of training in open warfare, some of it with tanks. Special attention was paid to conditioning, for the troops had to make approach marches for several nights before the battle, occupy the line and then make a deep penetration on foot carrying heavy loads. After it was over, excitement over the early successes, disappointment and bewilderment over the command failure to reinforce and exploit them, disgust at the collapse on the right flank in the face of the Germans' counteroffensive, and pride at the gallant resistance on the Bourlon flank were some of the reactions of the soldiers who took part in an experience so varied and original that it left its mark on all of them. It was the first epic battle since the 1st Ypres and the Somme. Soldiers could see their enemy, manoeuvre against him in small units, and measure their successes in ground gained. Those who took part had seen for a moment the light at the end of the tunnel. They now had a formula for ending the murderous attritional warfare directed by staff officers far in the rear. The new artillery technique had astounded the enemy; the tanks felt that they had proved themselves and confounded doubting infantrymen. With better cooperation with the tanks, and with more and better tanks, the Germans could be beaten. Surprise had been restored to the battlefield.

At GHQ and Army headquarters, the promise of Cambrai was overshadowed by the ominous implications of its failures and by a return to the defensive. The ease with which the German counter-blow had sliced into the flank of Third Army, the collapse in Russia, the manpower crisis, and the maneuvering of Lloyd George against Haig's autonomy combined to make defensive theory, not attack, the topic of the hour. In this context Cambrai taught the higher commanders and the lower ones rather different lessons. Originally intended only as a tank raid to disrupt and destroy communications centers, rout the front zone defenses and break up the artillery, it became an attempt at a cavalry breakthrough, which was Haig's persistent dream, despite the absence of infantry or tank reserves to exploit success. It resulted in tired and weakened units holding a vulnerable salient, as had so many other offensives. One flank of the salient had then been smashed in by an attack unheralded by the usual long bombardment. Clearly, with tanks, either side could smash through the forward defenses, but would have trouble maintaining momentum, not only for want of infantry reserves but also for lack of fresh tank battalions. In addition, the new style of attack depended on artillery intelligence which took time to collect before each fresh phase in operations.

Thus "limited operations" were still *de rigeur*, but they took a new form and purpose after Cambrai. They could achieve surprise because the new artillery plan could be prepared before the guns arrived and because the infantry-tank teams could penetrate quickly. On the other hand, the endurance of the tanks and infantry limited the depth of the penetration and a pause was required to gather new artillery intelligence. Each battle was attritional because it was designed to inflict higher casualties on the defenders than on the attackers and to force the defender to move his reserves for fear of a breakthrough. But rather than a breakthrough, the aim was to unhinge the German defenses gradually by a series of surprise attacks at widely separated points.

The success of this strategy depended on the army being able to apply the lessons of Cambrai. First it had to learn how to defend itself against similar tactics and, within a few days of the German counterblow at Cambrai on 30 November, conferences were called to dispose the defenses of the Expeditionary Force in depth. As we have seen, the Germans had taken similar measures, albeit unevenly, during the 1916 and 1917 battles.[37] By the end of 1917 they were convinced that no variations in the location and size of counterattack forces or in the distribution of defenders in the various zones could ensure that the forward zone would be held in the face of the new artillery methods and tank tactics, unless the ground and the weather were unusually favorable. Cambrai had shown that even the main battle zone might be overrun, in which case the attackers would be vulnerable to nothing less than a planned counteroffensive. In the meantime, the defense in depth depended on infantry with very good morale, enterprising junior leadership unperturbed by isolation, and a high proportion of automatic weapons and mortars. That was also the recipe for effective attacking troops. But by the beginning of 1918 there could be few units of such high caliber.

In evolving their new defense tactics, the Germans had gone more than half-way towards defeating their enemy when he used similar ones. Reinforced by divisions and command staffs with mobile experience on the Eastern Front and granted some months to train for the Michael offensive in March

1918, the Germans created *Stosstruppen* to penetrate the British defense in depth. Armed with light and heavy machine guns, mortars, grenades and rifles, they were to dribble forward between British-defended localities and to use column tactics to penetrate to gun positions, road junctions, and headquarters rather than to seize forward defenses. Heavy columns of infantry with artillery would follow to deal with obstinate defenders holding important ground. It was expected that the defenders' resistance would be uncoordinated and would collapse when road junctions in their rear were under direct fire, their communication with headquarters was broken, and they were attacked from flanks and rear. The scenario was similar to that envisaged by J. F. C. Fuller for Cambrai. The battle would open with what was called a hurricane bombardment of gas, smoke and high explosive which would first be directed mainly at the guns, the headquarters and the communications centers before turning into a bombardment of unprecedented intensity against the line infantry by mortars and guns of all calibers. Similar artillery plans had been successful at Riga and Caporetto. Here, the German technique varied from the new British system in two respects. First, as the Germans had no tanks, they bombarded the infantry before H-Hour; second, because their counter-battery fire relied on area bombardment rather than precision, and hence on neutralizing the guns by gassing and cutting their communications with the infantry rather than destroying them, they treated the guns, too, before H-Hour, and relied on the penetrating *Strosstruppen* to finish them.[38]

After Cambrai, British GHQ pressed armies to adopt a policy of defense in three zones of which the first was to be lightly held, the second the main battle zone, and the rear a base for counterattack forces. Continuous entrenchments were no longer to be dug, or, if dug, occupied. A proportion of the artillery was to remain silent until the battle started. Single guns were to be concealed in the forward zone as anti-tank weapons. Armoured telephone cables were to be buried extensively so that communications to isolated units would be ensured. Units were trained for mobile operations. The tanks were distributed along the front rather than being concentrated on the right where the German offensive was expected.

The success of the new tactics depended on adequate reserves and on divisions able to fight a semi-mobile battle with their flanks in the air. However, after the autumn battles commanders lacked reserves at all levels, for Haig's short period of relative independence from French strategy was over. He had had to send five divisions and General Plumer's staff to Italy, and had to take over the French front on the Somme and the Oise with six more. Petain seemed unconvinced that the German attack would come against the new British front, held now by Gough's Fifth Army, and against

Byng's Third Army on its left, and he avoided placing a reserve behind the British right before events showed it to be necessary. David Lloyd George had delayed replacing the casualties of the autumn at 3rd Ypres and Cambrai and the divisions had been reduced from twelve battalions to nine. Consequently, they were weak, needed rest and found the tasks of retraining, digging new defences and holding the front beyond their capacity. It is not surprising, then, that many senior commanders in both the British and the French armies were unconvinced that the new grand tactics of defense were sound when the reserves, the communications, the strong defenses and the rested and full-strength units were lacking. Nor were some of them sure that their soldiers understood the implications of the change that had taken place and would not fight better in continuous trenches. GHQ training memoranda, written in the usual cliches, and visits from the Field Marshal to explain the plan were not enough to ensure comprehension or compliance with tactics that had not been thoroughly examined. Consequently they were not uniformly applied.[39]

The German penetration on March 21st and succeeding days was aided by the relative weakness of the Fifth Army and by the thick morning mists that had also been a feature at Cambrai. Ludendorff's plan, much the same as that which had evolved in the British camp after Cambrai, was to mount a series of offensives at widely separated points to effect the gradual collapse of the front by exhausting British reserves. Ludendorff believed that he could rely on the French and British to separate in defense of the French capital and the channel ports respectively. But his initial deep penetration of the Fifth Army front led him to exploit with troops that he had nominated for his next attack on the Lys, and to leave them to guard his right flank, where the Third Army was relatively firm. His casualties were heavier than expected, and the exhaustion of the attacking troops, some of whom surrendered to the temptations of looting, was increased by the influenza epidemic that affected the Germans more than the Allies. Succeeding attacks on the Lys, the Aisne and the Marne achieved less and consumed the reserves he needed to retain the initiative. In July the tide turned. The Allies began a series of less ambitious operations that gradually loosened the front, first at one point and then at another, until the whole German Army was in retirement. Only autumn mud, broken roads, demolitions and the staunch rear-guard fighting of German machine gunners prevented the collapse of the German Army. The figures for prisoners taken in the final weeks indicate how close to disintegration it had come.

This brief survey does not claim to break new ground. Rather, it is an attempt to put generally known facts about the Expeditionary Force into a form of narrative that may be expanded and continued through the interwar years and the Second World War. A useful link between wartime and postwar military thought is the *Report* of the Kirke committee issued in 1932.[40] It was not widely distributed, probably because its conclusions were not in accordance with current policy. The contribution to it of Major General J. Kennedy casts an important light on the retrospective views of middle-piece commanders on the fighting in France and Belgium and provides a convenient conclusion to this essay.

Kennedy asserts that the commanders did not recognize three phases in an operation against entrenched troops: the break-in, a period of disorganized fighting later termed the dogfight by Sir Bernard Montgomery, and the breakout. Haig, for instance, recognised only the break-in and the breakthrough by the cavalry. Kennedy points out that a feature of the first phase was centralized and intricate staff planning; the second required divisional control of well trained teams of all arms; the final one required devolution of command and control functions and great initiative at sub-unit level. The chief failure of the British Army, Kennedy believed, was that having established a new command structure and learned new fighting techniques to meet trench conditions, the staffs and commanders were slow to devolve responsibility, and subordinates slow to assume it, to retrain the troops for the conditions of the dogfight and the breakout. The ad hoc affiliation of tanks with infantry and artillery was an obstacle to the growth of inter-arm cooperation and the shift of responsibility downwards. Principals did not know and trust one another, units had too little time to train together, and, in consequence, common doctrines were slow to evolve. Had the British established divisions with permanent affiliations, armoured divisions as the Kirke Committee conceived them, the results of the last hundred days might have been more decisive and more memorable.

Be that as it may, the events of the final hundred days did not impress themselves on the collective memory of the postwar British Army, whereas the Somme and "Passchendaele" could never be forgotten by the whole nation. Although the last days were the culmination of a learning process, very few had presided over the process from start to finish. Nor was it set down as a foundation for the doctrine of the future. Despite four years of fighting the British were as averse to doctrines of any kind as they had been before the war, when General Langlois commented that their manuals were good but that there was no uniformity about the way they were interpreted. "Sans doctrine, les textes ne sont rien. . . ." Consequently, when the commitments that had loomed so large before 1914 reappeared with peace and the memory of the last 100 days faded, there was no consensus about the immediate past, no history, and so, no doctrine. It is ironic that among the lessons least disputed was that *esprit de division* rather than regimental spirit had made the British Army so formidable.[41] For the creation of the division as a team had been the unfinished business of the General Staff in 1914. *Plus ça change. . . .*

Notes

1. T. H. E. Travers, "Future Warfare: H. G. Wells and British Military Theory," in Brian Bond and Ian Roy, eds., *War and Society* (London, 1975). Also Travers, "Technology, Tactics and Morale: Jean de Bloch, the Boer War, and British Military Theory, 1900-1914", *Journal of Modern History*, June, 1979.
2. Even Bond in his scholarly *The Victorian Army and the Staff College, 1854-1914* (London, 1972) does not quite reconcile the shortcomings of the commanders, the scarcity of the new staff officers and the high professionalism of the fighting units.
3. The militia myth reinforces the bias in favor of citizen soldiers. Another bias favors countrymen against townsmen. Yet, for instance, the Australian Expeditionary Force was not a force of leathery men from the Outback but predominantly from the cities. The press gave more publicity to the "Colonials" than to the British, so that it appeared that the latter were not only fighting to the last Frenchman but to the last Colonial too. Colonial divisions did become elite, but not at once and partly because they were treated more considerately. In June 1916 Haig commented in his diary on the failure of the 3rd Canadians at Hooge: "This seems bad and goes to prove that men with strange equipment and rugged countenances and beards are not all determined fighters." Haig Diary, 13 June 1916, typescript MS., National Library of Scotland.
4. The ferment of ideas displayed in the monthly *Royal United Services Journal* and the discussions at the annual General Staff Conference are evidence that the British Army was not amateurish. While the general run preferred sport to counting socks, a high proportion studied their profession even though they had not been to the Staff College.
5. Bond, *op. cit.*, p. 306. His point is that the senior commanders were either at the Staff College before the "Enlightenment" or were not "psc". His division of the Army into the "saved" and the "damned" is, perhaps, predetermined by his subject, and reflects the modern idea that the fount of all wisdom is the Staff. However, he is right to stress the small supply of staff-trained officers and the shortness of time in which the Staff College influence was at work as negative factors.
6. Commissioned conscripts and territorials entered the army with a profound suspicion of regulars and politicians and many had radical views about social and economic reform.
7. A description by John Terraine.
8. John P. Campbell, "Refighting Britain's Great Patriotic War," *The International Journal*, Vol. 26, 1970/1, 686. A useful review of the historiography of the two wars presenting a case for treating the period 1914-45 as an era, like that of 1789-1815, and with less help from hindsight than hitherto.
9. Campbell, *op. cit.*, p. 703, for several attempts to write them although "historians have been content to reconnoitre rather than penetrate the war's tactical core."
10. Maxse Papers, Box 1, Imperial War Museum; Staff College, Camberley, *Minutes* of the General Staff Conferences (G. S.) for the period. A fuller account of the preparation of the Expeditionary Force was presented in a paper, Graham, "The Development of Tactics and Weapons . . . 1907-1914" at the Canadian Historical Association, June, 1977.
11. Public Record Office (P.R.O.) London, W.O. 32/6745, 32/6679, 163/20 and 106/296; *Royal United Services Journal (RUSJ)*, Vol. 51, 1907, 362; Vol. 58, March and April

1914, in "Military Notes."
12. P.R.O., W.O. 105/46, April 1911 and 105/47, n.d., for a comparison with German conscripts. A. J. A. Wright, "The Probable Effects of Compulsory Service...," *RUSI*, Vol. 55, December 1911, describes British recruit material.
13. *RUSI*, Vol. 58, March 1914, in "Military Notes." The estimated effective cost of the army, discounting 1 million pounds spent on aviation, was to be less in 1914/15 than in 1907/8.
14. Examples are the manufacture of the MK III rifle, the reduction of the RAMC in peacetime, the delayed replacement of the Maxim and the restriction of the horse establishment in the Royal Garrison Artillery, in all of which economy was the deciding factor. Yet, considering that 1914 was not known to be the date when the war would start and that the British Army was not committed to the Continent, Treasury policy was sensible.
15. Discussion of the rival tactical systems was widespread. An example is *RUSI*, Vol. 51, 1907, "Military Notes," 632.
16. G.S.C., 1911, discussed the French system, of which the artillery was rightly suspicious. *RUSI*, Vol. 52, March 1908, and *Journal of Royal Artillery (RAJ)*, Vol. 38, July 1911.
17. At least 233 battalions would have received machine gun companies by the end of 1913. *RUSI*, Vol. 57, March 1913, 363; and Vol. 58, May 1914, 619.
18. G.S.C. of 1910 and 1912; G. S. Tweedie, "The Call for Higher Efficiency at Musketry," *RUSI*, Vol. 58, May 1914, 652; A. H. C. Kearsey, "The Manner in which the Infantry Attack can be best Supported...," *RUSI*, Vol. 54, 1910, 753.
19. G.S.C., 1912. *Memorandum on Army Training, 1912*, seen in the Maxse Papers; Sir James Edmonds, "The Conception and Birth of Some of the R.E. War Babies, 1914-18," *The Royal Engineers Journal*, Vol. 57, 1944.
20. G.S.C., 1910.
21. There is extensive literature on the artillery debate. Much information is scattered through the pages of Sir John Headlam's *The History of the Royal Artillery, Vol. II, 1899-1914*. Captain B. Vincent, "Artillery in the Manchurian Campaign," *RUSI*, Vol. 52, 1908, 28, gives a representative view of the lessons of that war. For the infantry aversion to and misunderstanding about indirect shooting, see G.S.C. of 1911. On the unwarranted independence of battery commanders, see *Memorandum on Army Training, 1912*, in the Maxse Papers.
22. "N", "The Organization of a Division," *RUSI*, Vol. 57, 1913, 1163. The theory is that reconstruction has taken the line of least resistance in that divisional commanders have left the integration of the artillery and infantry until last because they have the strongest institutions.
23. Headlam, *op. cit.*, p. 258, and *passim* explains the underprivileged status of the siege and heavy artillery. But the Garrison regarded the Field as bow and arrow gunners and the latter thought the Garrison were slow and academic. General Rowan Robertson, manuscript, "Guns in the Great War," Royal Artillery Library, Woolwich.
24. Weapons are neutral in the struggle between attack and defense. It is the use to which they are put that matters. The lack of fresh soldiers and the courage to risk what they had against an indomitable enemy had defeated the Germans. Had the fourteen British Territorial Divisions been trained and equipped in October 1914, the German front would have been broken at 1st Ypres.
25. There was very little liaison between the training staff at home and the divisions at the front. The former were always a step behind.
26. "So during 1916 we had the remarkable state of affairs where the largest army we had

ever produced would have been completely outshot by the archers of Crecy at 300 yards. . . ." W. D. Croft, *Three Years with the 9th (Scottish) Division* (London, 1919), p. 18.
27. My principal source for artillery matters during the war is Brigadier E. Anstey's draft, "History of the Royal Artillery in France and Belgium, 1914-18," at the Royal Artillery Library, Woolwich.
28. Many divisions never worked with tanks throughout the war. Others were skeptical of their value and resisted the methods advocated by the Tank Corps. The 51st Division at Flesquieres is an example.
29. Unless otherwise stated, I have used *Military Operations: France and Belgium*, compiled by Brigadier General J. E. Edmonds, as a principal source of battle details. For Neuve Chapelle, I have also used P.R.O., W.O. 95/2, Sir Douglas Haig's report; 95/706-8, Narratives of Operations IV Corps and G.S.; 95/630, II Corps papers on Neuve Chapelle; Haig Diary, March 1915; John Baynes, *Morale: A Study of Men and Courage* (London, 1967).
30. P.R.O., W.O. 32/3116, "Report of the Committee on the Lessons of the Great War," Appendix II, October 1932. Major General J. Kennedy remarks that the "Battle of the Broken Front" with broken units and broken communications was still neglected in FSR II, Ch. VII, when he was writing. It was still called the "Break Through" which assured that it was not dealt with as a separate phase from the "Break Out" and "Pursuit".
31. Haig Diary, 5, 7 and 13 April 1916; P.R.O., W.O. 95/431, Fourth Army; 95/673; III Corps; 95/851 X Corps; 95/895 XIII Corps; 95/922 XV Corps; A. H. Farrar-Hockley, *The Somme* (London, 1964).
32. Rawlinson's night move and dawn attack on 14 July, 1916, along the lines that he practised before the war, showed that his tactics were not unimaginative. General Monash is usually credited with perfecting the corps set-piece battle but he stood on the shoulders of many others.
33. I have relied for my view of German tactical development on G. C. Wynne, *If Germany Attacks* (London, 1940) and on the translations and summaries of German publications issued by GHQ during the war. The ground did not favor German defense theory at Arras, Vimy or Messines.
34. Major General H. Uniacke was Gough's Gunner. He was an exponent of the heavy bombardment and of the mobility of artillery, but the first prevented the second at Ypres.
35. Haig Diary, 14 October and 2 November, 1917.
    Brent Wilson, a graduate student at the University of New Brunswick, has drawn my attention to the possibility that Haig, being a late convert to the heavy preliminary bombardment, was slow to change his views a second time.
    Cambrai was an embarrassment to Haig. It made him appear to have refused French help while complaining that he lacked troops to exploit his success. It demonstrated the futility of continuing 3rd Ypres after September.
36. Haig Diary, December, 1917.
37. G. C. Wynne writes that von Lossberg's defence scheme met with resistance and was unevenly applied but was accepted in principle by Ludendorff as a result of the Somme. The Hindenburg line was designed according to his specifications. Wynne, *op. cit.*, 158-161.
38. Wynne and Anstey, *op. cit.*
39. A. H. Farrar-Hockley, *Goughie* (London, 1975); Joseph Gies, *Crisis 1918* (New York, 1974).
    It seems from the descriptions in some divisional histories and in Ernst Junger, *The*

Storm of Steel, that the new German attack methods were not applied uniformly either.
40. P.R.O., W.O. 32/3116.
41. Croft, *op. cit.*, p. 21.

TECHNOLOGY AND TACTICS

Aircraft versus Armor: Cambrai to Yom Kippur

Brereton Greenhous
Canadian Department of National Defence

The ideal way to use aircraft against tanks is to strike at their soft skinned and necessarily large logistical "tail". A fifty ton tank with a 120 mm gun and every kind of sophisticated aiming device is only a fifty ton pile of steel without gas or ammunition. But it is not always possible to achieve ideal solutions in war, an occupation which has been aptly defined as one of improvisation under pressure. It may therefore be advisable—under certain circumstances essential—to use aircraft directly against armour.

Although during the First World War both Great Britain and Germany developed specialized armored aircraft and comparatively sophisticated doctrines for close air support of ground forces, they were primarily concerned with attacking infantry and artillery.[1] The problem of aircraft in an anti-armor role was to find a way to add to the low-flying airplane's inherent speed and flexibility an adequate degree of immunity from ground fire, to combine with it the precise hitting power needed to destroy or disable a tank, and to do both these things without losing too much of the machine's ability to survive in a hostile air environment. Bombs, which must either hit the tank or, if big enough, land very close to it, could not be delivered with the requisite accuracy and the machine gun with which contemporary aircraft were usually equipped lacked hitting power.

First World War aircraft were quite capable of carrying tank-stopping guns, however. The Allies developed large calibre airborne cannon capable of knocking out any contemporary tank with one shot. Even as early as the fall of 1915 French Voisins were equipped with a 37 mm Hotchkiss intended for use against both air and ground targets, but these slow, ungainly pushers were far too vulnerable to ground fire or attack from the rear by enemy fighters. They were never employed operationally.

More promising was the Hispano-Suiza 1916 modification of a 37 mm Puteaux cannon so that it could be loaded with one hand during flight. It was mounted between the "V-eed" cylinder blocks of a Hispano-Suiza 220 hp engine and fired through a hollow airscrew shaft. This *moteur-cannon* was installed in two specially designed Spad fighters. The French aces Guynemer and Fonck claimed to have shot down a number of German aircraft with it. The arrangement impressed Cecil Lewis sufficiently for him to mention it in his superb First World War memoir, *Sagittarius Rising*, but before German tanks finally appeared on the battlefield in the spring of 1918 in very small numbers the concept of the "big gun" fighter seems to have disappeared.

What about the Germans, to whom Allied tanks posed a very distinct threat? The 20 mm machine guns that they fitted on a limited number of their ground attack machines in 1918 were capable of penetrating contemporary tank armor,[2] but they never seem to have given serious thought to using them in that manner. The *schlachtfliegern* were generally seen as an offensive rather than a defensive weapon. Very few of them were deployed against the British attack at Cambrai, for example. But when the Germans counterattacked ten days later and recovered with infantry nearly all the ground that the British had won with massed armor, their battle squadrons played a major role.

Allied accounts of enemy aircraft attacking tanks are rare. That was only to be expected when one bears in mind the noise, limited visibility and general disorientation that beset tank crews in action. The only references to any attack by German aircraft on British armor that I have been able to find relates to "ditched" tanks: on 31 July 1917, the notorious first day of 3rd Ypres (Passchendaele), when a tank officer noted that "many ditched tanks, lying well behind our infantry, were attacked by the enemy's pilots, who machine-gunned the crews. . . ."[3] On the German side, at Cambrai on November 21, 1917, rain and low clouds over the battlefield led the commander of the only German fighter flight in the area to report that "as aerial combat was still impossible, we carried out the tasks normally done by infantry pilots; our own front lines were reported on constantly and ground targets—especially tanks—were fired upon."[4]

At Amiens (8 August, 1918):

> On the orders of the Air Group Commander 23, Ground Attack Squadron 17 with seven aircraft started for the purpose of attacking enemy infantry, tanks and vehicle columns advancing from the direction of Villers-Bretonneux. Against these targets the squadron expended 3,630 rounds machine gun fire

and 725 bombs.[5]

As far as I can discover, none of the machines involved in these attacks had 20 mm cannon; any gunfire damage would probably have been limited to crew burns from bullet-splash. The 20 kg anti-personnel fragmentation bombs used at Amiens may or may not have been effective. A number of tanks were destroyed along the Villers-Bretonneux road that day but it is impossible now to distinguish which if any were victims of air attack.

A more specific and dramatic report from the second morning of the battle comes from the pen of Ernst Udet, who was then commanding the renowned Richthofen *Geschwader* in the "Flying Circus".

> We rise at day break. Tanks are coming!... A smoke screen is rising before them, and under its cover they creep across the flat, grassy plain. Fifteen of them, like mighty steel turtles.... They have already passed over the first German positions ... We dive, firing with both barrels, climb and come down again. No effect. It is like a woodpecker knocking against an iron door....
>
> The turtles continue to creep. One of them has now reached the dam [railway embankment]; it clumsily mounts the incline and rolls along the tracks.... I can now go at him from the side. Hardly three meters above the ground I go for him, go right up to him, hop over him, turn, and have another go. I get so close to him that I can make out every barrel and rivet on his steel plates. Even the washed-out clover leaf on his side, either a good luck charm or a unit insignia. I hop over him again, so low that my undercarriage almost touches the hump of his turret. I swing around and go at him again. Through the low-level tactic I neutralize the other tanks. If they fire on me, they fire on their own man.
>
> With the fifth attack I notice the first effect. Clumsily, the tank feels its way to the edge of the dam. It wants to get back down and into the covering smoke screen.... At the next moment he weaves and tumbles down the incline, landing on his back, helpless like a fallen bug.
>
> I come down from above, hammering my bursts into his lightly armoured underside. The tracks are still turning, the right snapping upward, reaching like the arm of a polyp and falling back. The tank now lies still, as if dead, but I still fire into it.
>
> A latch on the side of the gun turret opens. A man stumbles out, hands up in front of his bleeding face. I am so close that I can observe all this. But I can't fire any more, I have used up my

ammunition to the last round.[6]

Udet was flying a Fokker D VII, armed with twin synchronized 7.92 mm Maxim guns, usually (but mistakenly) called Spandaus by the Entente. The British Mk IV tank had a minimum of 6 mm of armour, so it is probable in this case that Udet was directly responsible for disabling the tank. No other case of a successful attack during the First World War has come to my attention.

During the interwar years there was little development of any kind in the realm of close air support. For most of that time the German Air Force was non-existent; and for all of it the Royal Air Force was obsessed by its own myth of strategic bombing in its determination to confirm for itself a totally independent existence. Even in the United States, where there was still an Army Air Corps (and where in 1919 experiments were made to study the effect of 37 mm airborne cannon fire against tanks)[7] there was progressively less concern with ground attack operations. By 1935, despite a nominal commitment to the concept testified to by the existence of eight "Attack" squadrons, the dominant idea of the Air Corps had come to be that "the possibility of simultaneously defeating a hostile air force and of attacking the enemy army in support of friendly ground forces was ... an alluring but false doctrine. . . . "The principal concern" of the Air Corps continued to be the development of long range bombers.[8] Of those air forces destined to play significant roles in the European and North African campaigns of the Second World War, only the Russians retained a clear though technologically weak commitment to the concept of close air support.[9]

When the German Air Force was re-formed as the *Luftwaffe* it envisaged "combat and other air action in support of the army forces on the ground" as one of its primary missions, but the terms *schlachtflieger* and *schlachtstaffel*—referring particularly to *close* air support—did not appear in its 1935 Operational Manual.[10] Interdiction bombing of one kind or another was still the Germans' primary tactical concern, the technological controversy revolving about the merits of medium level horizontal versus dive bombing. Only when Wolfram von Richthofen was replaced as Chief of Technical Services by the ebullient Ernst Udet in 1936 did the dive bombing concept begin to occupy a prominent place in official Luftwaffe thought. Even then the dive bomber was visualised primarily as the surest and most economical answer to the problem of "softening up" attacks on communications and vital rear areas, rather than as an instrument of close air support for the forward troops.[11]

The Spanish Civil War gave both the Luftwaffe and the Red Army Air Force a convenient proving ground on which to test their current air doctrines and technologies, but no deliberate policy of close air support was initiated by either at first. Only the obvious obsolescence, in their intended roles, of both the German Heinkel He 51 fighter and the Russian R-5 light bomber brought about their employment in low level, close support duties. The *Legion Kondor*—whose commander, von Richthofen, soon became an enthusiastic convert to the dive bomber concept in general, and to close air support in particular—experimented successfully with Henschel Hs 123 and Junkers Ju 87 dive bombers from 1937 onwards, while the Russians gradually replaced the R-5s with a less obsolescent machine, the I-15 single seater fighter. These aircraft were all used against artillery and infantry of uncertain morale and discipline, and they enjoyed considerable success in a low-intensity air environment and a ground environment of only moderate hostility.[12]

Aircraft were used against armor on rare occasions, most notably at Guadalajara and Teruel; but there has been no adequate study of their effectiveness and more research needs to be done in this area. A German work published in 1938 claims that "it was the air force that was the fastest moving and acting anti-tank weapon. . . . Such low-level attacks are usually carried out by single-seater fighter planes that are fully adequate."[13] But one German pilot of such machines certainly did not find them adequate. In his autobiography, Adolf Galland reports that:

> Our armament and equipment were at that time relatively primitive. We flew mainly with radio: it was regarded as a displaceable luxury. . . . Our HE 51's carried six 10-kg splinter bombs inside the fuselage, and were armed with two machine-guns, which had to be reloaded by hand after each burst. . . .[14]

In the Far East a largely ignored but major campaign (ignored because it was not accompanied by any formal declaration of war and was overshadowed by events in Europe) provided another indication of the ineffectiveness of aircraft against armor. In August 1939 large masses of Soviet and Japanese tanks came into conflict along the Mongolian border at Khalkin-Gol, or Nomanhan. Japanese aircraft attacked Russian armor but the Japanese assessed their own attacks as ineffectual in a post-campaign analysis. Soviet bombers and ground attack aircraft attacked the Japanese armor in flights of from 100 to 150 but did no damage, according to the leading Japanese authority. At a post-mortem on the campaign, the Japanese staffs decided that current technology was still inadequate to make the bomb-armed aeroplane an effective anti-tank weapon. They do not appear to have considered the possibility of airborne gun fire by something heavier than their standard 7.7 mm machine guns.[15]

In the Polish campaign of 1939 Ju 87 dive bombers were used mostly in the role for which they had originally been intended, to attack communications centers and troop concentrations behind the front. There were still problems of command and control in using close air support in the mobile battle—VHF radio was just becoming operational—and the *Stukas* flew mostly pre-planned missions against relatively static targets. The principal emphasis at the beginning of the campaign in the West was also upon interdiction but occasionally the *Stukas* were used in a close support role.[16] Against armor, they do not seem to have been very effective. Two examples must suffice. Outside Montcornet on 17 May 1940, thirteen light tanks of a regiment in the French 4th Armoured Division were caught "unprotected by air cover and desperately short of anti-aircraft guns" by "waves of Stukas." Only one of them was lost, according to the evidence of the tank company commander, whose account makes it clear that the division's counterattack on Montcornet had already failed before the Stukas arrived on the scene. Yet in 1962 a study prepared for the United States Air Force still accepted the conclusion of a 1949 German paper to the effect that the threat to Montcornet was largely averted by German dive bombers which put most of the French tanks out of action.[17] Again, on 21 May the brigadier commanding a British battle group counterattacking near Wailly found that for his 74 "Infantry" tanks "the main opposition came from his [Rommel's] field guns. . ., The air divebombing . . . did not worry the tanks much."[18] He lost only two of them to the dive bombers.

Nevertheless, from these campaigns in Poland and France emerged a myth of the dive bomber as an indispensable and devastating close support weapon that could destroy enemy armor almost routinely. It was driven home, emotionally rather than rationally, by newsreels depicting waves of Ju 87s peeling off in turn and hurtling groundwards to the accompaniment of shrieking sirens and howling engines, punctuated by the crump of bombs. Recognizing its psychological effect on ill-armed and unprepared troops (anti-aircraft defences in the Allied armies were minimal,[19]) no one attempted to define and evaluate its successes against armor in objective terms. By the end of June 1940 the besieged British were placing orders for dive bombers in America; in December the Vice Chief of the Air Staff (who certainly ought to have known better) was hounding his staff over delivery dates and noting that "Slessor and our technical people out in America strongly recommend us not to take other types instead of dive bombers"; and by the end of the year orders for 1,250 such machines had already been placed. But such was the panic and chaos caused by the sudden demand for dive bombers that, by 1942, the Air Ministry had no idea whether these orders had been cancelled, increased or decreased.

For the German failure to appreciate adequately the Stukas' limitations there was perhaps more excuse, since battles are rarely analysed as critically by the victors as they are by the vanquished. On the other hand, the wrecked or abandoned hulks of British and French tanks were available for the kind of elementary operational research which could have revealed a great deal concerning Stuka effectiveness against deployed armour. Had they drawn the right conclusions they might have turned more enthusiastically to alternative possibilities in the application of close ground support.

The Balkan campaign in the spring of 1941 turned out to be another "walkover" for the dive bomber, and in July the Germans marched against Russia with the myth of the Stuka further reinforced. The Red Air Force was decimated on the ground and the Soviet fighters that were left were no match for the Luftwaffe's Messerschmitt Me 109s. Unopposed Stukas enjoyed some success against the largely obsolete and lightly armoured Russian tanks that could often be disabled by the shrapnel of fragmentation bombs,[21] and everyone ignored the occasions when they failed dismally. On the fourth day of the invasion, for example, the whole of *Stukagruppe* 2 attacked a concentration of some 60 tanks fifty miles south of Grodno. It was later reported that only one tank had been knocked out, but the significance of that result was completely overlooked by the Germans in their euphoric advance. The commander of *Luftflotte* 2 has claimed that his men alone destroyed 1900 Russian tanks between June 22 and November 30, 1941, but it is impossible to check his figures and the enormous confusion that marked the early stages of *Barbarossa* make it very likely that his total is grossly inflated.[22]

However, within two months there appeared in large numbers two Russian harbingers of change, the T-34 tank and the Ilyushin Il-2 ground attack aircraft. The tank was proof against almost any airborne attack except a direct hit from a bomb or a 30 mm armor-piercing cannon shell from the rear at extremely close range, both very difficult feats to accomplish as late as 1944. The engine, fuel and crew of the Il-2 were practically immune to small caliber machine gun fire. A German air historian has recorded that "frequently formations of Me 109s and even Focke-Wulf FW 190s expended their entire allocations of ammunition firing at them without bringing them down."[23]

Some of these Il-2s were armed with RS 82 rockets, 82 millimeters in diameter and less than two feet long, weighing only 13.2 lbs. They were originally intended as an air-to-air fragmentation missile.[24] A Soviet pilot reported that "the rockets, of which each plane carried four [eight is the figure given by most secondary sources] and which shattered into innumerable fragments after covering a distance of 2,000 ft., were ... highly effective against ground targets",[25] but it would seem that he was referring only to soft-skinned vehicles and troops in the open, for the small 2 kg warhead and fragmentation effect were not compatible with an effective anti-tank function. For their original purpose the high degree of accuracy required of an acceptable anti-tank rocket was not essential, while the provision of it was the most technically difficult aspect of anti-tank rocketry. In fact, the Russians ofen preferred to use "rocket-bombs" (100 kg highexplosive bombs modified for armor piercing characteristics and fitted with 1 kg black powder charges in the tail) which achieved an impact velocity nearly double that of the conventional bomb, but which were certainly not precision weapons either.[26]

A later rocket, the RS 132, was an adaptation of the *Katyousha* ground-to-ground missile, approximately six feet long, five inches in diameter and fitted with armor-piercing and hollow charge warheads, but these missiles were intended to drench ground areas with fire and again the motor design was not modified for air-to-ground use. The Russians either did not try or failed to develop flight characteristics that would make the RS 132 into an effective anti-tank weapon, apparently preferring to rely on a "shotgun" approach. In 1942-1943 the commonest type of attack was as follows:

> Tanks were approached at altitudes varying between 2,000 to 2,640 feet from a steep gliding flight or a dive. First there was gun fire, the rockets were released at about 2,310 feet, then the bombs (preferably 220-pounders) were dropped, followed by cannon and machine gun fire, and then the aircraft pulled out of the dive at about 800 feet.[27]

Even with rockets released at the closer range, a one degree error in sighting on two axes—the Russians used a simple ring-sight focused at about 2,600 feet—would result in a "beaten zone" more than twenty-five times the area of a 10 × 20 ft tank. Such a probability factor, which takes no account of other variables such as aircraft yaw or "skid" at the time of release or the initial gravity drop of the rocket, was not likely to make rocket warheads of any size a dangerous threat to tanks unless fired in enormous quantities. This may explain the absence of reports from German tank crews on the Eastern Front recognising attacks by air-launched rockets, although (to give one example) the Russian 8th Army, fighting in the Don Basin in August 1943, claimed to have used "55,448 individual antitank bombs and 11,753 rockets" in the course of 15,642 sorties during that month. The accuracy of rocket delivery was probably even less than that of the anti-tank bombs and the likelihood of a "kill" was certainly much less.[28]

By Christmas 1941 the Germans were already becoming aware of their inability to stop tanks with bombers ("we could not, and did not do them any serious damage", recorded Kesselring in his *Memoirs*) and they had no air-to-ground rocket. It has been suggested, in a U.S. Air Force study, that "the inefficiency of the Russians in this regard actually retarded the Germans' appreciation of R[ocket] P[rojectiles] as a tank busting weapon".[29] The Germans foolishly neglected the advantages of their superior technology and sophisticated industrial base. In 1942 they showed little interest in air-to-ground rocket projectiles and turned to airborne cannon as their primary airborne anti-tank weapons.

They had two such weapon systems operational by the summer of 1943. On July 8, during the early stages of the decisive battle of Kursk (Operation *Zitadelle*) four squadrons of Henschel 129s, twin-engined close support machines armed with a single 30 mm cannon mounted in the nose, made a notably successful attack on a Russian armoured brigade near Belgorod. Their Mk 101 30 mm armor-piercing rounds with tungsten-carbide core could penetrate 80 mm of armor and, fired more or less at right angles, easily pierced the sloping 45 mm armor on the sides and rear of T34 tanks. But the Hs 129's maximum speed with the Mk 101 cannon pack was barely 200 mph and it was only possible to maintain the sustained attack needed to destroy the Russian tanks when a strong covering force of Fw 190s was present to suppress ground fire. Moreover, the Hs 129s were making their operational debut and the Russians were surprised and confused by the penetrative effect of the heavy cannon. Their tanks were concentrated and without air cover because previous attacks with bombs and small caliber machine guns had not been effective.[30] The Henschels were destined never to enjoy such success again.

During the same battle Hans-Ulrich Rudel, flying an experimental (and cumbersome) Ju 87 equipped with two 37 mm cannon (one under each wing) claimed twelve Soviet tanks in one day.[31] Before long Rudel was leading a wing of similar machines, but although they seem to have had a slight operational edge over the Hs 129s (probably because their engines were more reliable) neither type was more than marginally successful. They usually needed a close cover of fighter escorts. They were too slow and unwieldy to survive in anything but the most permissive environment. In mid-1943 the Germans were enjoying their last taste of anything more than a very brief and local air superiority and the already dangerous Russian "flak" was becoming ever more deadly.[32] Rudel, it is true, was to claim a personal score of 510 tank kills by the end of the war, but Rudel was noordinary man.

However, to all such claims a *caveat* is necessary. Operational research was an Anglo-American concept not seriously applied to tactical air operations until late 1943. Figures of tanks destroyed should not be taken at anything like face value. The appearance of smoke or flames on or around a tank could be deceptive, and, even when it was not, tanks in battle are very likely to be under fire from several sources at once, each of which may claim a kill. Heavy anti-tank or field guns were likely to do much more damage than airborne weapons. So were the various kinds of infantry anti-tank rocket weapons when used at short range. Airmen (and sometimes, perhaps, air historians!) have tended to attribute much too high a proportion of armor destroyed in battle to air power's unaided efforts. Such generalized claims as Plocher's that "hundreds of Soviet tanks were destroyed by aerial gunfire" or the Soviet Official History reporting their aircraft as having inflicted "great losses"[33] on German tank concentrations should be discounted almost entirely.

After *Zitadelle* the Germans were on the defensive in the East. "The vast open battlefields . . . enhanced the value of anti-tank aircraft which alone could deal adequately with tank formations which had broken through,"[34] and by December 1944 the Eastern Front had stretched to a length of 1,900 miles. But slow, unwieldy aircraft such as the Ju 87 and the Hs 129 were not the most suitable machines for such work in an environment increasingly filled with Soviet fighters and multiple automatic 20 and 30 mm flak. In October 1943 the Germans at last began to phase out the Ju 87. The Hs 129s were built in very limited numbers, less than a thousand altogether, and losses more or less equalled production; they remained a marginal weapon system.[35] The Russians still relied primarily on the Il-2, which was only slightly more successful than the German machines in the face of automatic cannon fire. However, they produced enormous numbers of them: more than 36,000 during the course of the war and 5,000 of their successors, the Il-10, according to the Russian official record.[36]

In the autumn of 1943 both Russians and Germans began to introduce fighters armed with various types of 30 or 37 mm cannon for anti-tank work, the Russians using a variant of the Yak 9 and the Germans a modified FW 190. Both machines were fast enough at low-level to be "in and out" often before the enemy flak could focus on them, but the weight and bulk of their cannon made them too unwieldy to look after themselves in air combat. They were easy pickings for air superiority fighters. Nor were even 37 mm cannon entirely satisfactory against the armor of the latest tanks, and although both types were, during the last months of the war, equipped with rockets, neither Yak nor FW carried projectiles with the power and accuracy needed to make them the deadly threat to heavy armor that was evolving in the West. The anti-tank rockets fitted to the FW 190, even after October 1944, were merely simple adaptations of the *Wehrmacht's Panzerschrecht* ground-to-ground missile (*Panzerblitz* I) with too light a warhead (1.5 lbs), too short a range (100 metres), and a requirement that the launch vehicle reduce speed to a maximum of 300 mph during release.[37]

In the United Kingdom, by mid-1942 even Churchill and Beaverbrook had finally been convinced of the dive bombers' inadequacy,[38] and the Air Staff's automatic rejection of anything that might be interpreted as an insidious attempt to establish an Army Air Arm was being slowly eroded by the realities of war.[39] The first British fighter-bomber had become operational in November 1941, and their first fighters equipped with 40 mm cannon for a specific anti-tank role during the summer of 1942 in the Western Desert of North Africa. In training, these aptly nicknamed "canopeners" had achieved an accuracy of 60% or better and their guns had proved capable of destroying or disabling contemporary tanks, but their numbers were so small (there were never more than three squadrons of them, rarely more than nine machines operational in each squadron) and their losses so heavy (averaging one out of three pilots and two out of three aircraft per mission during the desert campaign) [40] that they played no significant part in the North African war. Von Melenthin, who was very aware of air power, makes no particular reference to them, although he recognizes the growing strength of the Desert Air Force and the importance of the air arm in the open desert battlefields of 1942-3.[41]

Meanwhile, in July 1941, the Ordnance Board had suggested that rockets might provide a means of permitting aircraft to attack both shipping and tanks with some prospect of success; in August the Russians demonstrated their airborne rockets to British observers; and in October first trials had been carried out in Britain using Hurricane aircraft and a modified 3-inch anti-aircraft rocket motor.[42] By May 1942 it had been found possible to mount eight such rockets on Hurricanes and Swordfish, and in December, 1942 an Air Ministry report concluded that rocket-firing aircraft were "capable of making very accurate and devastating attacks".[43]

The British had originally planned on high-explosive warheads for anti-shipping work and solid, armor-piercing ones for anti-tank duties, but early experimentation led them to reverse this. Although the rockets had a dispersion factor of about half a degree (so that in a perfectly executed attack only half of them would land within ten yards of the center of the target when fired at the recommended contact range of 1000 yards),[44] the physical shock equivalent to eight six-inch shells landing in the immediate vicinity was supplemented by a substantial impact on morale, while a direct hit, even if it did not kill the crew, was almost certain to disable them, physically or psychologically, to the extent that they would be harmless for some time to come. Consequently, "by the end of 1942 the rocket projectile . . . was considered more suitable for the anti-tank role than the 40 mm gun,"[45] undoubtedly to the great relief of the men who flew the "can-openers". Production of Hurricane IID's was stopped and shortly afterwards the remaining aircraft were shipped to Burma, where they did notable work against a variety of ground targets in what had become a very permissive, low-intensity environment.

The attachment of just four rockets and their launching rails, however, pulled the Hurricanes' cruising speed down to 200 mph[46] and made them extremely susceptible to any kind of automatic flak before launching their missiles. Even after the rockets were fired and the rails jettisoned the Hurricane was too slow to look after itself in air-to-air combat. But once again the British were muddling through. The Hurricane's intended successor as an air superiority fighter had a disappointing rate of climb and lacked the performance at higher altitude to make it a success in that role. On the other hand, it turned out to have a speed of close to 400 mph at low levels, good ground visibility while in the air and, eventually, a construction capable of absorbing a great deal of punishment. In the summer of 1943 the Typhoon and the 3-inch rocket projectile were brought together to form the first really adequate anti-armor airborne weapons systems. (It still lacked, however, the modification of the Mk IID gyroscopic gunsight, introduced in late 1944,[47] which allowed not only for the initial gravity drop of the rocket but also for the influence of wind on the aircraft and for any movement of the target, thus making it possible to aim these powerful weapons with higher precision.)

When the Allies landed in Normandy there were ten squadrons of rocket-armed Typhoons of the British 2nd Tactical Air Force (TAF) in support and, even without the MK IID sight, the machines played a vital role during the first two months of the campaign on both the British and American fronts. On the ground the superior armor and armament of the enemy's Panther and Tiger tanks compensated for their numerical inferiority and, even allowing for the general Allied air superiority which made it so difficult for the Germans to bring their armor forward, the security of the bridgehead could have been seriously endangered by any of several attempted armored thrusts.

In this situation the rocket armed Typhoons, with their battlefield mobility and overall flexibility, were the most effective anti-tank weapon, breaking up concentrations before they could even launch their attacks.

Said a Canadian eyewitness, Peter Simonds:

> One rocket was enough to blow a tank to smithereens. I saw one Panther in the Tilly area whose turret and 88 mm. gun [sic] had been hit and blown about forty yards away from the hull of the tank. Five other enemy tanks were standing nearby—all apparently destroyed by rockets.... At a hamlet called Lignorolles, about a mile and a half west of Tilly, there were eight burned-out Panther tanks within an area smaller than a football field. They had been given a Typhoon massage....[48]

On the first occasion that the Panzers were used en masse in an offensive role, in northwest Europe when von Kluge launched his counterattack with elements of five armored divisions against the Americans at Mortain on August 7, the Typhoons stopped the Germans in their tracks. "American tactical aircraft at this time were not equipped with effective rocket armament for attacking armor", so the Thunderbolt fighter-bombers of the U.S. IX Air Force attacked transport and communications while the Typhoons concentrated on tanks.[49] In a morning mist 2 Panzer Division, spearheading the attack,

> made a swift advance of about ten miles and suffered only three tank losses.... Suddenly the Allied fighter-bombers swooped out of the sky. They came down in hundreds firing their rockets. ... We could do nothing against them, and we could make no further progress. The next day the planes came down again. We were forced to give up the ground we had gained, and by 9 August the division was back where it started from north of Mortain, having lost thirty tanks and 800 men.[50]

2 Panzer had begun the attack with about eighty tanks; more than one-third of its battle strength appears to have been destroyed by the Typhoons in the course of 294 sorties. Later the commanding officer of the 9th U.S. Division's 39th Infantry Regiment, which was split in two by the initial German assault at Mortain, is said to have observed that "our anti-tank force was insufficient to deal with such a force of Panther tanks.... Although in general he considered that claims from the Air Corps [sic] on tanks were optimistic, he did not disbelieve a claim of the order of 100 for the total effort that day.[51]"

The first American squadron armed with the 5-inch HVAR [High Velocity Aircraft Rockets], four per P47 Thunderbolt, only became operational on 15 July and played a minor role in the Mortain battle. Both their 5-inch and the British 3-inch rocket had a dispersion factor of about 10 mills. at 1000 yards, but the Typhoons carried twice the number of rockets and the British were far more expert at delivering them. Every British pilot had had a three week course in rocket-firing (with regular refresher training) but the Americans had had only "limited training—limited both in time and in number of rockets fired. In fact, one or two pilots had not fired a rocket at all before they took off that day."[52] Their wartime commander has commented that in 1944 "I don't like to admit this but it is true, that there was an attitude that went all through the Air Force, that I adopted, my juniors adopted and my seniors adopted, that this [close air support] was not our mission. It was not our mission to participate that close in battle." When asked what type of theater training was given to new or replacement pilots in IX TAC he answered, "Practically none. . . . They had a lot of time in the airplane. They had a lot of air-to-air gunnery. What they didn't have was air-to-ground work. They weren't very well trained on that."[53]

When the Americans attacked armor directly, they usually relied upon fighter-bombers which, even at low level, could not match the accuracy of rocket-armed fighters. A report prepared by the German Air Historical Branch on Allied close support during June 1944 states categorically that "the rockets scored more direct hits, even from greater heights, than the normal fighter-bomber attacks."[54]

The overwhelming and ever-expanding Allied air superiority in the West made it increasingly difficult for the Germans to concentrate any quantity of armor. After their counterattack at Mortain the Typhoons had few occasions to demonstrate their capability on any scale. One example is to be found in the After Action report of the United States' 2nd Armored Division for 26 December 1944, the climactic day of the German Ardennes offensive. The division "directed air attacks which knocked out an additional 18 tanks". This engagement is also reported on in a manuscript memoir of Major General E. H. Harman.

> General White knew the British on the Meuse had some rocket-firing Typhoons in their area and, while our artillery attempted to delay the advance, he hustled over to Dinant. The British were most anxious to cooperate, but now a new difficulty presented itself. Second Armored had no radio capable of communicating with and guiding the planes. White had an answer; our artillery observation Cubs had precise knowledge of the enemy's whereabouts.
>
> A little later, American soldiers were treated to an odd spectacle. A squadron of Typhoons appeared in the sky led by one

tiny, armored Cub. It was like a butterfly leading a squadron of buzzards. The Cub dived on Tiger and Panther tanks coming up the road towards Celles and then made tracks towards our lines. The Typhoons screamed down, rockets sizzling from their wing runners, and left devastation in their wake. [On this day 2nd TAF claimed only two tanks destroyed by seven Typhoons firing 53 rockets.][55]

By the end of the campaign in North West Europe the value of tactical air power in combating armor had been firmly established. However, the successful employment of the atomic bomb at Hiroshima enabled British and American air leaders to revitalize the largely discredited theory of strategic bombing as a decisive weapon, and they abandoned many of the tactical air concepts that events had virtually forced them to adopt in the last half of the Second World War. When the Korean War began and the North Koreans started south, equipped with Russian T-34 tanks, although the Americans easily established air superiority over the battlefield there were no air units equipped with or trained in the techniques of anti-armor operations.[56] There was also a grave shortage of ground-based anti-tank equipment, and what field guns there were failed to make much impact on the North Korean T-34s. During July 1950 these tanks spearheaded the Communist drive down the Korean peninsula. F-80s armed with 5-inch HVARs did their best to combat them but peacetime training had been severely limited. "Since few FEAF [Far East Air Force] pilots had ever fired a 5-inch HVAR, they would have to get their rocketry training in the heat of combat."[57]

Moreover, the F-80 interceptors were not really suitable for the fighter-bomber role, and HVARs were not the best kind of rocket to use against the well-sloped armour of the T-34s. The high-explosive head tended to ricochet off and the debris thrown up by the blast often damaged the F-80 so that had fired it as much as the tank it was aimed at.[58] Fortunately for the Americans, however, the T-34s had rubber-tired bogies and a hull that was susceptible to napalm. Once pilots had learned to aim napalm tanks accurately they were able to knock them out.[59] Marine Corps air units (F4U Corsairs), which, with no strategic bombing doctrines to distract them, had retained their Second World War expertise, also helped stop the North Koreans, who were brought to a halt outside Pusan (in what was virtually a last stand), through the decimation of their armor from the air when they were within a few miles of driving the Americans into the sea. The destruction of 80 of the 300 tanks with which the North Koreans had started their attack was credited to tactical air power.[60]

A 6.5 inch shaped charge head for the HVAR appeared soon afterwards, based on the shaped charge for the 3.5 inch ground-to-ground anti-tank rocket that by a fortunate coincidence had gone into production in the States just fifteen days before the North Koreans attacked.[61] This ATAR [Anti-Tank Aerial Rocket] was first used in September 1950; by the end of October those North Korean tanks that had not been destroyed had been reduced to operating singly or in pairs, moving only at night. An Operational Research Study issued by the American Army's Far East Command in February 1951, based on operations between July 1950 and January 1951, reported what happened to 329 T-34s.[61] Of the total 165 tank kills, 102 (62%) were the result of air attack and 63 (38%) were attributable to ground attack. The method of air attack broke down as follows:

Napalm accounted for	60 (59%)
Rockets	18 (17%)
Bombs	7 (7%)
Strafing (50 cal and 20 mm)	7 (7%)
Not specified	10 (10%)

In Korea, however, the power of airborne anti-tank weapon systems was inflated by the weakness of the ground arm during those early days. Not until the Arab-Israeli Six Day War of 1967 were analysts able to make a satisfactory comparison between ground and air power in the anti-armor role. The Israeli Air Force did a great deal of damage in Egyptian and Syrian armor after the Egyptian Air Force had been destroyed on the ground. Among the 527 Egyptian armored vehicles examined by Israeli operational researchers in the Sinai, ground-based weapons systems accounted for 195 (37%), air force 80 (15%), 203 (39%) were abandoned and 49 (9%) were "unexplained". Nearly all of the vehicles destroyed by air action were caught on the Sinai tracks in the course of the general retreat towards the canal. There were three main lines of retreat taken by the Egyptians, a northern one along the so-called "Jerusalem road" and two southern ones either funnelling down the Wadi Giddi or over the Mitla Pass. The bulk of the damage inflicted along the Jerusalem road was done by Israeli ground forces, 51.3% of the enemy armor being destroyed by ground fire and 7.5% by air attack. On the two southern routes, however, ground fire accounted for 11.87% and airborne fire for 26.97%, suggesting that the air attack in the southern and more inaccessible part of the Sinai was a good deal more intense than in the north. Total Egyptian armoured losses on the northern line amounted to 265 vehicles, compared with 204 in the south. It may therefore be an indication of the moral effect of air attack that in the south 60% of the Egyptian armor had been abandoned for one reason or another, as compared with only 24% in the north. (The remaining 58 vehicles were widely spread in the desert, mostly upon the east bank of the Canal.)[63]

In the initial major engagement between ground forces, in the vicinity of Rafah where the Egyptian tanks were dug in, only one Egyptian tank, an old T-34, seems to have been disabled by air action. Twenty-seven others, mostly Josef Stalins, were destroyed by ground forces and thirteen were abandoned without being hit. But dug-in armor is an unsuitable target for close support aircraft (although the Second World War indicated that carpet bombing by heavy bombers, such as accompanied the launching of Operation *Cobra* before the St. Lo breakout in Normandy, can effectively immobilize it). In these initial attacks only light and comparatively lightly armed Fouga Magister training aircraft were used in the close support role, while the Mirages and Mysteres were engaged against SAM sites and in interdiction duties.[64]

Once the Egyptian rout had begun the higher-performance aircraft were able on occasion to take time off from their interdiction role to play a part in the close air support function. A flight commander with a Mystere IV squadron flew twenty such missions and "five were against tanks—new tanks—and the rest of them against artillery or APC (Armored Personnel Carriers)". On one occasion he led his four Mysteres against fifteen or sixteen T-54s that had ambushed Israeli AMX 13 light reconnaissance tanks at the mouth of the Mitla Pass.

> We had two napalms per aeroplane and a 30 mm gun which had armour-piercing and incendiary ammunition. I told my followers: "Look, the situation is bad here. We are going to do it like on the range; not like in a war, get rid of everything and go home, but one by one and as many passes as we can." We hit all the sixteen tanks, six of them with napalm and the rest of them with the 30 mm, and all of them had their fuel tanks on. The T-54 had jettisonable fuel tanks on the deck, so we saw them hit with gunfire. . . . we hit all the sixteen and we got contact with another flight that was going the same way and brought them in as well. . . . I still remember that one of my lads didn't hit with his first napalm, and I told him, "Look, I'm going to send you back home if that's the performance you're going to show here," and then somebody else didn't hit one but all the other six napalm bombs that we had on the flight hit the tanks and then with the 30 mm we had five passes.

These attacks were flown at a speed of "400 knots" from an approach of about 30 degrees, with cannon fire beginning at a range of 1000 yards at a height of 500-600 feet.[65] Israeli operational researchers confirmed twelve kills in their subsequent examination of the battlefield. The others may have been destroyed by ground fire.[66]

The battle for Jerusalem brought about what may well have been the most dramatic and effective use of aircraft against armor to date. A Jordanian armoured brigade en route from Jericho to Jerusalem on the evening of 6 June was caught by IAF air strikes in the vicinity of Edom Ascent. According to the IAF historian the Israelis flew about 300 sorties, using napalm, rockets, bombs and cannon fire. Five aircraft were lost and all the Jordanian tanks were destroyed, primarily by 2.75-inch rockets, or abandoned. No exact statistics can be established, however, since the Jordanian tanks were American Pattons, a type also used by the Israeli armored forces, and all those which could move or be readily repaired were taken over at once by the Israelis.[67]

One of the most interesting aspects of the aircraft-armor conflict has been the time lag in developing an adequate means of combating close air support with surface-to-air missiles (SAM) and 20-40 mm automatic flak. These types of defences only began to play a significant part in the evolution of tactical air operations, especially close support functions, in the latter stages of the Vietnam conflict, a development that led the Americans to adopt an aircraft designed specifically for close support work in the form of the Fairchild A-10. Here is an account by an American airman of an early SAM encounter:

> We got a SAM warning, I looked up and saw three of them high, and just about the time Rocky started to break I saw two more coming low. The two low ones passed behind us, the three high started to come down on us.... I heard a call coming from Covey 282. They were hit about two miles off our left wing—he just flat went down.... Rocky broke to the right but a SAM was coming up from the tail.... the SAM went off in full view of the canopy at about 300 feet [range].[68]

In order to keep clear of mobile SAMs it was sometimes necessary to use less traditional forms of airborne support. In April 1972, when North Vietnamese armor was moving south, B-52 heavy bombers were used to attack a column of enemy tanks on Highway One by means of pattern bombing reminiscent of Operation *Cobra* in Normandy during the Second World War:

> the column was moving between two pre-planned targets, so we went on secure voice and started it back through Task Force

Alpha to 7th Air Force. In about 30 minutes 6 "Bufs" [B-52s] came in and rippled that road. They got 35 tanks and, we found out later, the command bunker of the NVA division in that area.[69]

More commonly, fighter-bombers were still used and if the enemy targets were not defended by SAMs, then the "slow-movers"—propeller-driven, sub-sonic machines—were effective. A Marine colonel recorded that when "a flight of our VNAF A-1 aircraft came on station, within two hours 13 of 16 T-54 tanks had been destroyed by mines, tank fire, air strikes and Vietnamese Marine Infantry weapons. Bombs were typically delivered in a shallow, 15 degree dive and released about 800 feet above ground, but "when accuracy was less critical or ground fire more intense, pilots made dive bombing attacks from angles of about 45 degrees, releasing (officially) from 7000 feet.[70]

According to the authors of this semi-official but unsubstantiated monograph, on 28 April 1972, when the North Vietnamese launched a "tank and infantry attack on the north end of the bridge leading into Quang Tri City. . . . The attack was beaten off [by Tactical Air] and resulted in 5 out of 5 tanks destroyed to the northwest of the bridge."[71] But one wishes that there was something more than USAF claims and reports to base these figures on.

In really close country, helicopters armed with guided, rocket-propelled missiles and using "nap of the earth" tactics to approach enemy armor were sometimes effective. At An Loc, where a dozen T-54s and PT-76s got into the city, anti-tank helicopters destroyed three with rockets in the streets and drove the others into positions where they were easily handled by ground-based, light anti-tank weapons.[72] Another new weapon in the armory of the airborne anti-tank force was the "smart" or laser-guided bomb which offered a new and phenomenal level of precision.

> I found two tanks just north of the Marines position on the My Chanh River. It was at twilight. There was a PT-76 and a T-54. The PT-76 was trying to pull the T-54 out of a dry stream bed. I called for ordnance and there was none available. I waited and finally Schlitz and Raccoon, two F-4s out of Ubon, showed up. They were equipped with a laser-guided bomb system known as Paveway One. Raccoon was the "illuminator," that is, he carried the laser gun used to direct the laser energy onto the target. Schlitz carried the laser-guided bombs.
>
> They checked in with two or three minutes of "playtime" left—that is, they were running short of fuel. I briefed them on

the way in to save time. I put the smoke down marking the target. By this time, the illuminator, Raccoon was in orbit, he asked me which tank we wanted to hit first. I suggested the one that was not stuck. Within about 30 seconds he said "I've started the music" meaning the laser beam was on the target. Schlitz was already in position for the drop—the LGB hit right on that PT-76, blew the turret off and flipped the tank over. The blast covered the second tank with mud, so I put another smoke rocket down, Raccoon "started the music" again. Schlitz meanwhile had pulled right back up on the porch for another run. The whole operation was over in three minutes. Two bombs—two tanks destroyed. I logged them in at 6:18 and off at 6:21—that must be close to a record of some kind.[73]

American casualties in Vietnam were not heavy, for the enemy's anti-aircraft resources were severely limited and there was no counter-air of any significance. The Arab-Israeli Yom Kippur War was the first high-intensity conflict involving SAMs and radar-predicted, multiple flak guns in the anti-aircraft defence. The attackers, on the other hand, had improved conventional munitions such as anti-armor cluster bombs and an air-to-surface rocket salvo that could now deliver fifty or more rockets per aircraft in a shotgun type pattern. However, when armor had the protection of an established anti-aircraft screen, the cost-effectiveness balance seems to have swung back in favour of the ground forces. When, on 6 October 1973, the IAF "intervened in full strength" to check Syrian tanks on the Golan, they lost thirty machines even though there was no serious opposition from the Syrian air force.[74] In the course of the war the Israelis lost 109 aircraft, 40 to SAMs, 31 to AA fire, 21 in air-to-air combat, two to friendly forces and 15 to unknown causes. Most of the 15 attributed to unknown causes were probably lost to ground-based fire. (236 more suffered battle damage; 215 "were repaired in less than one week." Of the other 21, a number were doubtless damaged beyond repair, most by ground-based fire, and could be added to the total number lost.)[75]

There seems to be no hard evidence to tell us how effective Israeli airpower was in the anti-armor role. Most disabled vehicles had received multiple hits for which Israeli operations research analysts were unable to assign precise responsibility. On the Golan, by accepting heavy casualties in order to play a part in resolving a crucial situation, the IAF certainly had some effect, but a German tank officer who visited the scene in early November 1973 estimated that 80% of Syrian losses (about 1300 tanks) could be directly attributed to Israeli tank fire "whereas only 20% were destroyed by the effects of artillery or air forces."[76] Israeli writers have observed that "the aircraft hit Syrian tanks but couldn't stem the flood...."[77] Initially, IAF loss rates ran at 10-14% of close-support sorties flown; subsequently this dropped to approximately 8%.[78] This was hardly acceptable (although it may have been necessary) and it looks even worse when matched up against the rare (and perhaps atypical) account of specific damage done.

> At 15:00 hours the Syrian tanks began to roll across the trench [according to an Israeli tank squad commander.] "They passed alongside the position—not in front of it.... There were 20-30 tanks that crossed the bridges. Within a few minutes, Air Force planes appeared and hit a Syrian tank on the bridge...."[79]

The Syrians could do no better. Four Mig 17s, firing rockets, attacked Israeli tanks "from about 60 metres up.... They killed one [Israeli] tank and two of its crew—their only hit."[80] On both sides the ground-based anti-air threat apparently affected pilot performances to a significant extent.

In the Sinai, too, anti-armor air was not cost-effective until, on 14 October, the Egyptians made the mistake of launching an assault that took them out from under their anti-aircraft umbrella. On their Third Army front major attacks were directed at the Mitla and Giddi Passes, and an armored brigade was ordered south from Port Ibrahim towards the Abu Rodeis oil fields on the northern shore of the Gulf of Suez. This last was a totally unexpected thrust and, "since there were not enough IDF armoured forces available to intercept, the IAF was called in." The Egyptians had moved beyond their fixed missile screen.

> For four days IAF aircraft attacked the Egyptian armour and managed to stop it, inflicting heavy losses in armoured vehicles and finally compelling the enemy forces to withdraw. Whenever Egyptian armour tried to move again in this direction it was attacked from the air until finally the Egyptians gave up all further attempts. After the war heavy damage to Egyptian armour was confirmed. It was obvious that here IAF alone managed to stop the enemy armour advance, using standard armament from high [7000 feet] altitude. No aircraft were lost.[81]

Air power, using improved conventional munitions and precision guided missiles, can now dominate the battlefield and provide an effective antidote to rampaging armor as long as air superiority can be retained. However, since air superiority now involves surface-to-air guided or homing missiles and radar-predicted multiple automatic flak as well as enemy fighters, the value of close support aircraft will be greater where ground based anti-aircraft systems can be most effectively suppressed. It seems that this will be in a defensive environment, where an enemy break-in (or even break-out) has meant that his tactical light anti-aircraft systems may be caught off balance because they have not yet been able to site themselves to best advantage, and where the heavier systems are either not yet present or are not dug in. Under any other circumstances the massive intervention is not likely to be cost-effective, even when the air arm has succeeded in establishing a comprehensive local superiority over the rival air force. But cost-effectiveness is not always what matters in war. There will be times, after an enemy breakthrough has occurred, when the enemy must be stopped. Tactical air power, because of its flexibility and battlefield mobility, will then be invaluable while it lasts.

Notes

1. For a more specific examination of close ground support in the First World War, see B. Greenhous, "Close Ground Support Aircraft in World War I: The Counter Anti-Tank Role" *Aerospace Historian*, XXVII, (1974); and "The Evolution of a Close Ground Support Role for Aircraft in the First World War", *Military Affairs*, XXXIX (1975). See also J. R. Cuneo, "Preparations of German Attack Aviation for the Offensive of March, 1918", *Military Affairs*, VII (1943), 69.
2. A maximum of 14 mm on British tanks, although the German A7V carried 30 mm of frontal armor. Until the late 1930's armor-piercing ammunition was, as a rule of thumb, capable of penetrating armor plate equal in thickness to the caliber of the weapon from which it was fired at an angle of 30° or more. During the Second World War the penetrative capability of solid shot was slightly more than doubled by the introduction of higher velocities and harder and heavier cores. Penetration fell off significantly as range increased beyond 50 metres, however. See E. J. Hoffschmidt, *German Aircraft Guns and Cannon* (Old Greenwich, Conn.: WE Publications, 1969).
3. D. G. Browne, *The Tank In Action* (Edinburgh: Blackwood, 1920), 124.
4. R. Flashar, "In der Tankschlacht von Cambrai," in G. P. Neumann (ed), *In der Luft Unbesieght* (Munich: J. P. Lehmans, 1932), 99.
5. *Die Luftstreitkrafte in der Abwehrschlacht Zwischen Somme und Oise vom 8 bis 12 August 1918....* (Berlin: E. S. Mittler & Sohne, 1942), 161.
6. Ernst Udet, *Ace of the Iron Cross* (New York: Doubleday, 1970), pp. 83-84.
7. Major W. B. Hardigg to Major Walton, 24 March 1922 (Enclosure, "Firing Against Tanks and Armour") in Air Corps Control File 1917-1938, 473.5/Gen, RG 18, US National Archives, Washington, D.C.
8. R. F. Futrell, "Ideas, Concepts, Doctrines: A History of Basic Thinking in the United States Air Force" (Air University: Aerospace Studies Institute, 1971), 72 and 75 (Unpublished MS).
9. Harder Diary, 20 July 1933 to 13 August 1933, in K.-H. Volker (ed.), *Dokuments and Dokumentenfotos zu Geschichte de deutschen Luftwaffe* (Stuttgart: Deutsche Verlag-Arstalt, 1966), 87-91. Attempts to produce a specialized ground-attack machine began in 1930 but no successful designed appeared until 1938. See also G. le Q. Martel, *An Outspoken Soldier* (London: Sifton Praed, 1949), XIV, for a first hand account of a Russian demonstration in 1936.
10. USAF in Europe, "Air Staff Post-Hostilities Intelligence Requirements on G.A.F. (Tac-

tical Employment, Fighter Operations, Vol. I)," microfilm serial 2905, National Archives.
11. P. Deichmann, *German Air Force Operations in Support of the Army* [USAF Historical Studies, No. 163] (New York: Arno Press, 1962), 35; Werner Baumbach, *The Life and Death of the Luftwaffe* (New York: Ballantine Books, 1967), 8; R. Suchenwirth, *Historical Turning Points in the German Air Force War Effort* [USAF Historical Studies No. 189] (New York: Arno Press, 1968), 28-31.
12. For a variety of first hand accounts, see Albert Kropp (ed), *So Kaempfen deutsche Soldaten* (Berlin: Limpert, 1939); Wulf Bley (ed), *Das Buch der Spanienflieger* (Leipzig: von Mase under Kohler, 1939); Didier Poulain, "Aircraft and Mechanized Land Warfare: The Battle of Guadalajara, 1937", *Journal of the Royal United Service Institute*, LXXXIII (1938), 362. One of the accounts in Kropp is by Harro Harder, who had studied his part in Russia four years earlier (no. 7 above). A German summary of the Russian experience is given in D. W. Schwabedissen, *The Russian Air Force in the Eyes of German Commanders* [USAF Historical Studies, No. 175] (New York: Arno Press, 1960), 47-8.
13. F. A. Fischer and E. A. Billeb, *Luftmacht* (Heidelberg/Berlin: Vowinckel, 1938), 86
14. Adolf Galland, *The First and The Last* (London: Methuen, 1955), 29-30.
15. See A. Ikuta, *Japanese Army Air Operations in Manchuria* (Japanese National Defence College: Asagumo News Press, 1976) [Translated from the Japanese]. See also Ikuta to Greenhous, 3 October 1975, Greenhous Papers, Directorate of History, Department of National Defence, Ottawa (DHist).
16. Deichmann, 156; Anon., *The Diary of a Staff Officer* (London: Methuen, 1941), 11.
17. Alistair Horne, *To Lose a Battle: France 1940* (Boston and Toronto: Little, Brown, 1969), 429; Deichmann, 148.
18. Quoted in B. H. Liddell Hart, *The Tanks*, II (London: Cassell, 1954), 14.
19. "In 1939, when the Luftwaffe possessed seventy-two anti-aircraft regiments, France had only five, and was noticeably short in the small calibre 25 mm and 40 mm guns essential for protecting ground troops from attack...." Horne, 84-5.
20. Freeman to Haining, 16 December 1940. The chaotic circumstances of the dive bomber question in Britain are far from clear in the Air Ministry files, and no study of the British aircraft purchasing arrangements has yet analysed it adequately. The relevant text and this reference summarize, to some extent, correspondence in Air 19/233, Public Record Office, London.
21. K. Rokossovsky, *A Soldier's Duty* (Moscow: Progress Publications, 1970), 14.
22. Deichmann, 48; H.-U. Rudel, *Stuka Pilot* (Dublin: Euphorian Books, 1952), 88; A. Kesselring, *Memoirs* (London: William Kimber, 1953), 97.
23. K. Uebe, *Russian Reactions to German Air Power in World War II* [USAF Historical Studies No. 176] (New York: Arno Press, 1964), 7.
24. W. von Braun and F. I. Ordway, *History of Rocketry and Space Travel* (New York: Crowell, 1966), 86. RS 82s had first been used operationally in the 1939 Mongolian campaign against the Japanese. At the time the Russians claimed a number of victories for them, including two on the day that they were first used (5 August 1939), but post war research indicates that on the day in question the Japanese suffered no losses. Eichuro Sekagawan, "The Undeclared Air War", *Air Enthusiast*, IV (July 1973), 26-7. There is no record that the Japanese ever realised rockets were being used, and the Soviet Official History now simply records that "Rockets were used on fighters as early as the battles at Khalkin-Gol." R. Wagner (ed), *The Soviet Air Force in World War II* [Soviet Official History, translated by Leland Fetzer] (New York: Doubleday, 1973), 15.
25. Schwabedissen, 104.

26. M. Caidin, *Rockets and Missiles: Past and Future* (New York: McBride, 1954), 98-99.
27. Schwabedissen, 221 and 347.
28. In 1943 Soviet dive bombers could expect to score 64% hits on a target 200 metres square. Wagner, 193.
29. Kesselring, 97; USAF in Europe, microfilm serials 2930-2931.
30. Deichmann, 148-49; William Green, *Warplanes of the Third Reich* (New York: Doubleday, 1970), 397; and Cajus Bekker, *The Luftwaffe War Diaries* (New York: Doubleday, 1968), 297.
31. Rudel, 86.
32. Wagner, 215; Rudel, 86-87.
33. H. Plocher, *The German Air Force Versus Russia, 1943* [USAF Historical Studies No. 155] (New York: Arno Press, 1967), 99; Wagner, 353.
34. Suchenwirth, 107.
35. Deichmann, 33-34; Cajus Bekker, *The Luftwaffe War Diaries* (New York: Doubleday, 1968), 376.
36. Wagner, 12.
37. *Ibid.*; H. J. Nowarra, *The Focke-Wulf 190: A Famous German Fighter* (Letchworth: Harleyford, 1965), 154-155. The Panzerblitz II, with a greatly increased range and a 4.5 lb hollow-charge warhead, appeared too late and in too small numbers to play a significant role.
38. Arthur Tedder reported on the action at Bir Hacheim in June 1942: "Heavy losses inflicted on *Stukas* attacking Hakeim [*sic*]. . . . French suffered practically no losses from *Stuka* attacks." A year later, when the British knew that the Germans knew it too, and the world was being informed that "the dive bomber is now dead," Britain's South-East Asia Command was puzzling over the "publicity problem" posed by the first of those 2,000 American dive bombers, which had just become operational in the Far East. Air 19/233, PRO.
39. The RAF side of the controversy can be found in the Air Historical Board's *Air Support* [Air Publication 3235] (London: Air Ministry, 1955). For the army's case, not made as strongly as it might be, see C. E. Carrington, "Army/Air Co-operation, 1939-1943," *Journal of the Royal United Service Institute*, CXV (December 1970), 37-41. Until a more detailed and objective study appears we might agree with Sholto Douglas that the RAF's Army Co-operation Command was a "miasma of frustration." Douglas to Freeman, 11 August 1942, in Air 8/984, PRO.
40. Biographic File, Sqdn. Ldr. A. J. Simpson, DFC, DHist. NDHQ, Ottawa.
41. Von Mellenthin, 119-20.
42. A. F. S. Napier, "British Rockets In The World War," *Journal of the Royal Artillery*, LXIII (1946), 11-15. Prewar and early wartime research had emphasized the rocket as an anti-aircraft weapon. See also A. D. Crowe, "The Rocket As A Weapon of War in the British Forces," *Journal of the Institute of Mechanical Engineers*, CLVIII (1948).
43. Crowe, 17.
44. Alfred Price, "The 3-inch Rocket: How Effective Was it Against the German Tanks in Normandy?" *Royal Air Force Quarterly* (1975), 127-28.
45. G. F. Wallace, *The Guns of the Royal Air Force 1939-1945* (London: William Kimber, 1972), 135.
46. P. Clostermann, *The Big Show* (London: Penguin, 1958), 76.
47. M. M. Postan, D. Hay, J. D. Scott, *Design and Development of Weapons* [British Official History] (London: HMSO, 1964), 119.
48. Peter Simonds, *Maple Leaf Up, Maple Leaf Down* (New York: Island Press, 1964),

179-180. The Panther was armed with a 75 mm. gun; only the Jagdpanther was equipped with an 88, and it had no turret.
49. C. F. Shore, *2nd Tactical Air Force* (Reading, Berks: Osprey, 1970), 18; L. F. Ellis, *Victory In The West* [British Official History] (London: HMSO, 1962), I, 414.
50. Von Luttwitz, quoted in M. Shulman, *Defeat in the West* (New York: Ballantine Books, 1968), 197. Von Luttwitz, commanding 2 Panzer, remembers that the morning "dawned bright and clear," but all other accounts say otherwise.
51. M. Blumenson, *Breakout and Pursuit* [U.S. Official History[(Washington: Department of the Army, 1961), 474; 83 Group 2nd TAF Intelligence Summaries Nos. 55 and 63, Air 25/704, PRO.
52. John E. Burchard (ed), *Rockets, Guns and Targets* ["Science in World War II" Series, Office of Scientific Research and Development] (Boston: Little, Brown, 1948), 170.
53. "Interview with Gen. Elwood R. Quesada, 12 May 1975," USAF Oral History Tape K 239.0512-838, transcript in US Army Military History Collection, Carlisle Barracks, Carlisle, Pa., 64 and 143.
54. 8th Abteilung Report, 27 August 1944, in DHist. 981.013 (D33).
55. Ms memoir and "Military Correspondence, September 1944-June 1945" in Harman Papers, US Army Military History Collection, Carlisle Barracks, Carlisle, Pa.
56. W. M. Reid, "Tactical Air In Limited War," *Air University Quarterly Review*, VIII (1956), 43.
57. R. F. Futrell, *The USAF In Korea, 1950-1953* (New York: Duell, Sloan and Pearce, 1961), 58.
58. *Ibid.*, 84.
59. R. F. Futrell, "Air War In Korea: II," *Air University Quarterly Review*, IV, No. 3 (Spring, 1951), 62.
60. *Armor*, LIX (1950), 8.
61. *Ordnance*, XXXV (1950), 201 and 213.
62. ORO-R-3 (FEC) *Close Air Support Operations in Korea* (GHQ, Far East Command, February 1951), 337-38.
63. Israeli Operational Research map overlay, 1:250,000, in Wallach Correspondence, Greenhous Papers, DHist. Israeli pilots, in the heat of action, claimed hits on 476 armoured vehicles. If one half of those vehicles abandoned, were abandoned as a result of air action, then the ratio of claims to kills is approximately 3 to 1. It should be noted also that Dr. Steven Canby's figure of only about 5% of destroyed tanks showing signs of being attacked from the air and only 2% being seriously damaged [*Adelphi Paper*, 109 (Winter 1974/75) p. 37] is wrong.
64. *Ibid*; see also "Nahost-Luftkrieg im July 1967," *Wehrkunde*, XVI (1967), 452.
65. Author's interview with Colonel Avrihu Ben-Nun at Hazor Air Base, Israel, 14 November 1978. Greenhous Papers, DHist.
66. IOR map overlay, *loc. cit.*
67. Wallach to Greenhous, 7 March 1979 and 12 October 1979, in Greenhous Papers, DHist.
68. A. H. C. Lavelle (ed.) *Airpower and the 1972 Spring Invasion* (USAF Southeast Asia Monograph Series, Vol. II, Monograph 3—no place or date of publication), 35.
69. *Ibid.*, 41.
70. *Ibid.*, 11, 47.
71. *Ibid.*, 49.
72. J. D. Howard, "An Loc", in *Army* (1975), 20.
73. Lavalle (ed.), *Airpower and the 1972 Spring Invasion*, 54.
74. E. Luttwak and D. Horowitz, *The Israeli Army* (London: Allen Lane, 1975), 374.

75. C. A. Olschner, "The Air Superiority Battle in the Middle East, 1967-1973: Part III, 1973" (unpublished dissertation for US Army Command and Staff College, 1978), 68.
76. H. A. Kiesewetter, "Panzerfriedhof", in *Wehrkunde*, No. 2 (February 1974), 75.
77. Yeshayahu Ben-Porat, *et. al.*, *Kippur* (Tel Aviv: Special Edition Publishers, n.d.—first published in Hebrew, 1973), 175.
78. C. A. Olschner, "The Air Superiority Battle in the Middle East, 1967-1973: Part III, 1973" (unpublished dissertation for US Army Command and Staff College, 1978), 49-50.
79. Quoted by S. L. A. Marshall, "Tank Warrior in the Golan", *Military Review*, Vol. LVI, No. 1 (January, 1976), 5.
80. Ben-Porat, *et. al.*, *Kippur*, 48.
81. Wallach to Greenhous, 7 March 1979 (citing IAF Historian), Greenhous Papers, DHist.

INTELLIGENCE

Secret Operations versus Secret Intelligence in World War II: The British Experience

David Stafford
University of Victoria

"If you have to work for a racket, let it be an old-established racket."[1,2]

There is a self-evident difference between the methods and objectives of secret intelligence gathering on the one hand and the covert promotion of sabotage and subversion on the other. Indeed, where operations of both kinds are carried out simultaneously in the same area, there is likely to be a direct conflict involved: "Intelligence, in the true sense of the word, is incompatible with violent, subversive, conspiratorial activity," notes one commentator,[3] while another, himself involved in both types of activity before and during World War II, has stated the conflict more directly: "the man who is interested in obtaining intelligence must have peace and quiet . . . but the man who has to carry out operations will produce loud noises if he is successful, and it is only too likely that some of the men he uses will not escape."[4] It is equally self-evident, therefore, that the closest coordination between these two kinds of activities is a pre-condition for their success and an essential safeguard against mutually self-destructive activity. Logic would seem to demand that while each sphere should have its own networks and agents, there should be one overall organisation and command to ensure coordination and cooperation and establish clear priorities. Logic, however, is rarely a reliable guide to practice, and the case of Britain in World War II is no exception.

Although there is still a widespread popular belief (reinforced by such egregiously misleading products of journalism as William Stevenson's *A Man Called Intrepid*) that such activities fell within the embrace of one all-powerful British secret service, in fact secret intelligence gathering on the one hand, and the promotion of sabotage, subversion, and guerrilla warfare on the other, were the responsibility between 1940 and 1945 of two separate organisations. The first function was assumed by the Secret Intelligence Service (SIS), sometimes referred to as M.I.6., and the second by the Special Operations Executive (SOE). The institutionalization of their differing and often conflicting functions was denoted by the fact that whereas SIS was responsible to the Foreign Secretary, Ministerial responsibility for SOE was vested in the Minister of Economic Warfare. The former organisation, which dated back to the period before World War I, had become a permanent if unacknowledged part of the machinery of British government, and its members considered themselves professionals. The latter, created only in the confused and desperate weeks following the collapse of France, was specifically designed, as was the responsible Ministry, to have no more than a wartime existence. Although many of its early personnel were drawn from SIS, its membership on the whole was of a different background and outlook, and SOE's opponents tended to regard them as dangerous amateurs.

Thus, to a functional conflict always potentially present was added a considerable degree of professional mistrust, analogous to, but quite distinct from, the troubled relationship between SIS and M.I.5 (domestic counter-espionage). Intensifying the mistrust was the fact that SOE had been created out of the merger of three existing secret departments, one of which was SIS's own Section D, created in 1938. The very existence of SOE, therefore, was a reminder to SIS of its loss of control over activities entrusted initially to itself. It is not surprising, therefore, that the SIS-SOE wartime relationship was often marked by considerable rivalry and hostility which went beyond the bounds of what might be described as legitimate conflict of interest, and that on at least one occasion SIS played a major part in attempting to destroy SOE as an independent agency. It is this conflict, and the ultimate outcome, which concerns us here.[5]

The SIS has been the focus of a considerable amount of recent attention. The revelations since 1974 about *Ultra* intelligence have not only opened up vast new areas for scholarship and initiated calls for a fundamental reevaluation of much of the war's strategic history, but they have also seen at least three attempts (conscious or otherwise) to rehabilitate the SIS's wartime image, which had been seriously tarnished by the public revelation in the mid-1960s that one of its brightest wartime stars, the head of its wartime anti-Soviet section, had been a Soviet agent. SIS's control of *Ultra*, the secret war winner, has cast the failure to detect Philby into the shadows.[6] SIS's pre-war activities have recently, and for the first time, come under the scrutiny of professional historians, and it is now known that apart from its strictly intelligence functions it sometimes intervened unwarrantably in other affairs. The researches of Christopher Andrew, for example,[7] have revealed that SIS was deeply implicated in the attempt to smear the Labour Party with the taint of Communism during the Zinoviev letter affair of 1924, and had earlier successfully destroyed sympathy for Sir Roger Casement in the United States by deliberately leaking passages from Casement's diaries dealing with his homosexual affairs. A seminar at the University of Leeds on "Secret Intelligence and Modern Politics" in January 1978 is a further evidence of increased interest by the historical profession in SIS affairs.

From these and other sources a picture has emerged of a Secret Intelligence Service whose formative and heroic years had been those of the struggle against Bolshevism, and which subsequently had great difficulty in freeing itself of anti-Bolshevik obsessions. In the 1920s the SIS was severely run down, and in the 1930s the SIS leadership, under Admiral Sir Hugh Sinclair, was more conscious of dangers from the Left than from the Right; hence there were no great difficulties for entry into the Service for Philby, who had carefully cultivated a right-wing image during the Spanish Civil War. Within the limits of its drastically reduced post-World War I budget, SIS relied upon a network of agents, usually with city or business connections, whose political and social sympathies and background produced a misappreciation of fascism and, at the leadership level, an identification with the policy of appeasement. Typically for an organisation devoted to the collection of intelligence, it distrusted intellectuals, and one of its major figures, Sir Claude Dansey, who headed its offensive espionage section, was proud to boast that "he would never willingly employ a university man." For these and other reasons there had been no clear vision of what the priority targets for intelligence should be, and Germany itself had not been accorded top priority. On the eve of war it was, according to one source, "run-down, unimaginative, and badly staffed."[8]

It was within this organisation that there was created in March 1938, shortly after the *Anschluss* and in response to it, "Section D", whose function was the planning of sabotage in Germany and Italy, the creation of lines of communication into these countries for the purpose of infiltrating gadgets and explosives for subversive use, the invention of sabotage devices and special wireless sets, and the composition and distribution of "black" propaganda. In March 1939, immediately following the German occupation of Prague, Section D was authorized to pass from the planning to the operational phase and actively to counter the Nazi threat in countries just occupied or threatened by Germany.[9] But Section D and its colorful head, Major Lawrence Grand (he was never to be seen without a red carnation, and Kim Philby, who became a member of Section D, wrote of him that his mind "ranged free and handsome over the whole field of his awesome responsibilities, never shirking from an idea, however big, or wild")[10], soon ran afoul of the regular SIS establishment. Even some of its own members who had been drawn from SIS were hostile to the idea of subversion,[11] and its many less than successful early attempts at sabotage and subversion led the SIS establishment to regard it as a Frankenstein's monster which would have to be destroyed before it irrevocably damaged SIS's own secret intelligence networks. In the Spring of 1940, however, Section D dug its own grave. In the confused and panic-stricken climate produced by the collapse of France and the Dunkirk evacuation, Section D had been given the task of organizing civil resistance and sabotage in the U.K. in the event of a German invasion. This became a fiasco: too many of its organizers were arrested as German agents while preparing ammunition dumps and expanding badger sets as underground HQs.[12] Despite some spectacular *coup de main* sorties in the face of German advances in Western Europe, Section D's days were numbered.

But if Sir Stewart Menzies, the new head of SIS, thought that the type of activities entrusted to Section D could be brought back under the firm control of SIS, he was wrong. SIS had had its own disasters, especially the Venlo incident, where, in November 1939, two of its leading agents in Europe had been led into a trap and captured by the Germans. SIS's entire European network was compromised by the Venlo affair. When Churchill took over control of the Government in May 1940 he quickly made clear his dissatisfaction both with SIS and the Intelligence services of the armed forces, which between them had failed to detect German plans for the invasion of Norway, by launching an enquiry into their operations, headed by Lord Hankey. Churchill was, moreover, inclined to place more, not less, stress on the importance of stimulating resistance through an active campaign of sabotage and subversion in occupied Europe, and, romantic that he was, envisaged an ultimate European uprising of the oppressed peoples contributing to the defeat of Nazism. As a direct result, in July 1940, Section D was merged with two other secret organizations, M.I.(R) of the War Office and E.H. of the Foreign Office,[13] to form a new organization, the Special Operations Executive. Control of SOE was bitterly contested within Whitehall corridors, but on July 16, 1940 Churchill entrusted it to Hugh Dalton, Minister for Economic Warfare and a member of the Labour Party's leadership, urging him, in Dalton's words, "to set Europe ablaze."[14]

SIS's previous relationship with Labour in power had been an unhappy experience, and Dalton's high flown rhetoric about SOE's role in Europe can only have fed their anxieties. References to creating movements in Europe like the Sinn Fein in Ireland and the Chinese guerrillas fighting the Japanese were seen as threats to SIS's need for the painstaking reconstruction of intelligence networks, and it is therefore not surprising that in its infancy SOE encountered considerable hostility from SIS.[15] The situation was not eased but was made substantially worse by the fact that SOE, while theoretically independent of SIS, was still heavily dependent upon it for personnel, communications, and various forms of technical assistance, all of which made it vulnerable to SIS pressure. As far as personnel went, many of its top executives were men with close SIS links. SOE's first executive head (denoted by the initials "CD") was Sir Frank Nelson, who had made a fortune in early life as a merchant in Bombay, and had worked for SIS since the 1920s while serving as Conservative Member of Parliament for Stroud. On the outbreak of war he was head of SIS's organization in Switzerland under cover of the post of consul-general in Basle,[16] and was appointed CD by Dalton to carry out a purge of Section D personnel. Assisted by some senior officers in SIS, his purge was thorough. According to Philby, "they were not only determined to 'get Grand', but to get all his closest henchmen too."[17] Grand was dismissed at a painful interview with Dalton on 22 August 1940, and he never forgave him for it.[18] Nelson's link with Dalton was maintained through Gladwyn Jebb, SOE's Chief Executive Officer (CEO). Jebb, too, had close links with SIS, for in his previous position as private secretary to Sir Alexander Cadogan, the Permanent Under-Secretary at the Foreign Office, he had been responsible for liaison between the Foreign Office and SIS. And SOE's director of intelligence and security after June 1941, Air Commodore Boyle, had directed Liaison between the Air Staff and SIS in the 1930s.

The presence of these and other individuals with SIS links was less, however, a cause of SOE's early dependence on SIS than a symptom of it. Of greater significance was SOE's dependence on SIS for its communications. Under an agreement reached between the two organisations in September 1940, all SOE's W/T (wireless) traffic was to be handled through SIS, which was given the right to accept or reject it. This dependence on SIS lasted until 1942, and gave rise to probably well-founded suspicions that SIS exerted its rights under the September 1940 agreement and often obstructed SOE operations of which it disapproved.[19] Colonel, later Brigadier, Gambier-Perry, the head of SIS communications, disliked and was alleged to have obstructed SOE, and the SOE-SIS rivalry of the early years of the war focused for the most part on this issue. Late in November 1940, for example, the communications situation "was regarded as so serious as to make CD's activities in many cases almost impossible," and when Dalton had a personal *tete-a-tete* with Mountbatten in January 1942, communications and the related rivalry with SIS formed the first topic of their discussion.[20] Further, under the same agreement of September 1940, all intelligence collected by SOE was to be passed to SIS, even before it was circulated within SOE.[21] Finally, SOE was to consult SIS in the recruitment of all agents. SIS, on the other hand, was bound by no reciprocal obligation, and made no commitment to provide SOE with any intelligence at all.

This a-symmetrical arrangement over intelligence, which lasted for the duration of the war, was crucial to the relationship between the two organisations and played a part in securing SIS's ultimate predominance. It meant, for example, that high grade information derived from Special Intelligence ('Ultra') was often not passed on to SOE at all, even when it concerned SOE immediately. Menzies mistrusted SOE's security and, anxious to safeguard Ultra, he rationed it parsimoniously and sometimes dangerously. On one occasion, for example, Section 17M of the Naval Intelligence Division (NID) noted from intercepts of *Abwehr* communications in Norway that the Germans had gone on alert along parts of the Norwegian coast. This information was passed by NID to SOE, in case the latter was planning to land agents ashore. This apparently sensible move incurred Menzies' wrath, however, for it breached a strict SIS rule whereby NID was not to pass Special Intelligence material to anyone outside the Navy. SIS alone would pass such information on; but, as NID was well aware, "the liaison between the appropriate one of "C's" officers and the "named officer" in SOE to whom some Special Intelligence was passed was "not too efficient."[22] For a long time, intercepts revealing the extent of partisan activity in Yugoslavia never reached SOE, and it seems to have been only by accident that these began to circulate to SOE Cairo at the beginning of 1943.[23] The justification for restricting access to Special Intelligence need not concern us here. The point was that the lack of reciprocity fed justifiable suspicions within SOE that SIS sometimes withheld important intelligence, or else released it at chosen moments to discredit or weaken SOE. This control of information was an invaluable asset in bureaucratic rivalry and SIS benefited enormously from it.[24]

    A further problem in SOE's attempts to free itself from SIS control and fulfill the independent tasks assigned to it by the War Cabinet arose in its relationship with the European governments in exile in London. The communication links in this area, too, were dominated by SIS, and in its liaison with their secret service department SOE was severely disadvantaged. The Europeans were often unwilling initially to deal with SOE, especially when they were warned against doing so by their long established SIS contacts. Things gradually improved here too, but as many or most of the European governments never established the clear administrative divisions analogous to the SOE/SIS demarcation of responsibility, there remained much ground for conflict and confusion.

    SOE had been created with ringing injunctions from Churchill and Dalton about the need to "set Europe ablaze." It quickly became apparent that, whatever the European factors involved might be, SOE was in no position between 1940 and 1942 to do so. The elementary tasks of creating an independently functioning administration, communications system, and operational networks abroad had to begin *ab initio*. Some of the obstacles we have noted. But a more profound factor impinged on SOE's relative subordination to SIS in the early years of the war. This was the priority given by the Chiefs of Staff to SIS over SOE interests in most of occupied Europe prior to 1944.

In the early days following the German occupation of Western Europe, British strategists and SOE itself had talked of resistance as an independent variable: a factor in the strategic equation which might autonomously determine the location and timing of British landings on the Continent, which would then "detonate" a European explosion against Nazism. Although remnants of this concept continued to grace Churchillian rhetoric until well into 1942, it had ceased to have any real meaning even before the end of 1940; by then it had become clear that SOE itself could not become operational until 1942 at the earliest, and that there was no hope of any British landings before that date. If the prime role of SOE was to prepare secret armies in Europe which would rise up only at the moment of such landings, then so long as landings could not be envisaged SOE's activities would have to take second priority to those of SIS, the source (so it was argued) of the vital intelligence needed for planning Continental operations. In other words, once British strategic planning had returned to an even keel after the spring of 1940 and it was recognized that allied landings would be determined in time and location by factors other than some vaguely defined concept of European resistance, the subordinate role of European resistance in Allied strategy became reflected in the subordination of SOE's needs to those of SIS.

That SOE's activities would be seen as a threat to SIS interests, and that the former would be subordinated to the latter, was made clear very early. In December 1940, for example, SIS opposed SOE activities in the coastal areas of northern and western France "as they might interfere with their organisation for getting agents into enemy-occupied territory"; and no SOE sea operations were ever attempted on the north French coast as a consequence.[25] No mention had been made in SOE's first directive from the Chiefs of Staff of September 1940 of any priority in SOE-SIS relations, but this is not surprising as SOE was not yet operational, and in any case was still totally dependent on and subordinate to SIS in the matters of communications. It was only in March 1942, therefore, the question became pressing, and it was rendered more acute because it was in the Spring of 1942 that Anglo-American plans for continental landings began to take shape.

It is clear from a wide range of evidence that SOE-SIS relations entered a state of crisis at this stage. Bickham Sweet-Escott recalls that Dalton's departure to the Board of Trade following the Government reshuffle of February 1942, and his replacement by Lord Selborne, "coincided with one of the periodical waves of feeling in Whitehall that it was time to give SOE a real shake up." In this SIS played a leading role.[27] In the course of the subsequent enquiry into SOE's affairs (known as the "Hanbury-Williams" or "Playfair" report), Sir Frank Nelson retired and was replaced by Sir Charles Hambro, whose many other responsibilities as director of the Bank of England and managing director of the Great Western Railway Company contributed to his ultimate dismissal and replacement by Major-General Colin Gubbins in September 1943. These personnel changes reflected profound turbulence between SOE and SIS at the operational level. In late April 1942 the Joint Intelligence Committee recommended to the Chiefs of Staff that SOE's activities be more strictly coordinated with and subordinated to the interests of SIS. In the words of the Committee:

(a) We recommend that closer coordination between SIS, SOE, and CCO [Combined Operations] be maintained at all stages in planning. The activities of SOE increase the alertness of the local authorities and greatly hamper the work of our intelligence. It is necessary that such activities should be avoided in areas where this adverse effect outweighs the results which these activities may be expected to produce. This applies especially to areas where larger operations are likely to take place.

(b) It is most difficult for the organisations concerned to determine by agreement where the balance of advantage lies, since their duties inevitably conflict. Until some adequate arrangements are made to eliminate this conflict, we recommend that the Chiefs of Staff should appoint someone to decide all such issues on their behalf.[28]

The Chiefs of Staff accepted the recommendations on 1 May 1942 and the Directors of Plans were instructed to examine and report on the best method of coordinating SOE and SIS activities.[29] The proposals made by the Directors of Plans remain classified, but shortly thereafter a liaison committee under Sir Findlater Stewart was set up to resolve SOE-SIS disputes. But it was short-lived, and subsequent coordination and liaison appears to have reverted to the ad hoc arrangements of the past, supplemented by regular meetings between SOE and the Foreign Office at which SIS was represented. Nonetheless, the principle that priority should be given to intelligence needs over SOE needs in the potential invasion areas had been recognised and affirmed by the Chiefs of Staff, and it determined policy throughout 1942 and 1943. SOE's 1942 Directive made no explicit reference to it; but this would have been otiose, given that it was issued almost simultaneously with the recommendations just discussed.[30] But the 1943 Directive of 20 March 1943 spelled SOE's subordination in very clear terms. The Chiefs of Staff informed SOE that:

You will continue to pass on all intelligence you may collect to

SIS. You may undertake the collection of intelligence for SIS in areas should SIS request you to do so.

The requirements of SIS should in general be accorded priority over your own operations in Norway, Sweden, France and the Low Countries, and, if the appropriate Commanders-in-Chief agree, on the mainland of Italy and in Sicily. In other areas great care should be taken that your activities do not clash with SIS and that the latter's sources of information are not imperiled.[31]

Thus, in Western Europe, SIS enjoyed officially sanctioned priority over SOE, and in most areas this extended into 1944 and up to the eve of the Normandy landings. It had always applied in Spain and Portugal, but did not, it is worth noting, apply to the Balkans, where in 1943 guerrilla warfare became an important adjunct to British strategy and to strategic deception.[32]

The March 1943 directive did not, however, solve the question of priorities. Rather, it was an indicator of a growing problem. The rising tempo of European resistance, the increasing demands being placed upon SOE from a number of sources, and the build-up for the 1944 invasion meant that 1943 saw an escalation in its activities. This escalation produced a demand for greatly increased resources, and in particular for aircraft for the dropping of supplies and infiltration of agents. In the summer of 1943 the Chiefs of Staff came under pressure from Churchill to supply more aircraft to SOE, a proposal which met with strong resistance from Bomber Command. The Chiefs of Staff argued that SOE efforts in Western Europe, where the secret armies would not be required until 1944, did not justify the increased numbers of aircraft for which SOE was pressing. In the course of high-level discussion on this issue, an incident occurred which illuminated graphically both the continuing depth of SOE-SIS rivalry and the way in which SIS attempted to use its control over intelligence and its influence with the military professionals to its own advantage against SOE.

On 17 June 1943 the Chiefs of Staff had recommended that the Air Ministry and SOE should jointly examine ways in which to increase aircraft deliveries in order to meet the expanded SOE targets, and a new dimension was added to the issue when on 23 June Churchill requested that supplies to the Balkan guerrillas be increased to 500 tons a month. The upshot was that on 24 July the Air Ministry reported that if SOE were to be given the aircraft to meet all its claimed needs, 70 aircraft would have to be diverted from the bomber offensive. To prevent this, and to scale down the size of the diversion, Portal, Chief of the Air Staff, argued that a distinction should be drawn between SOE work in the Balkans and that in Western Europe. In the Balkans, SOE activities should receive full support, as "they accord with our strategic plans, they exploit our present successes and should give us good and immediate results"[33]; but in Western Europe resistance movements were of potential value only and could not merit any diversion of aircraft until nearer the invasion date. Although SOE contested this view, the Chiefs of Staff, meeting on 27 July, accepted Portal's argument and agreed that "we should support SOE activities in the Balkans as far as possible and at the expense, if necessary, of supply to the resistance groups in western Europe." Their attitude cannot, however, have been uninfluenced by the sudden presentation by Portal of a Secret Intelligence report, dated only the previous day, which indicated that a number of resistance groups in France had been penetrated by the Germans. "At the present moment," this report noted, "resistance groups [in France] are at their lowest ebb and cannot be counted on as a serious factor unless and until they are rebuilt on a smaller and sounder basis."[34]

What better evidence could have been presented to reinforce the Air Ministry's case and to influence the outcome of the inevitable appeal which it knew SOE would make to the Defence Committee? That SIS and the Air Ministry had collaborated closely to produce the evidence at this opportune juncture seems clear; Bomber Command wished to prevent diversion of more aircraft to SOE, and SIS had always been hostile to SOE activities in France. SOE seems to have been given no time to comment on the allegations before they were produced by Portal. Given that the Joint Intelligence Committee subsequently declared the SIS conclusions to be misleading, this was a particularly serious omission.

The affair revealed graphically that SOE and SIS rivalry remained acute. This was so much the case that the Joint Intelligence Committee, on receiving instructions from the Chiefs of Staff to investigate the whole affair, decided that neither SIS nor the Ministry of Economic Warfare should take part in the investigation, although both were normally represented on the Committee. Having dismissed the general conclusions of the SIS report as misleading (although recognising that many groups in France had been penetrated), the Committee concluded that the most serious problems the affair revealed were the lack of any close SOE-SIS coordination, and the lack of symmetry in their respective relationships to the directing centres of British strategy: for whereas "C" (SIS) was, through permanent representation on the JIC, an integral part of the Chiefs of Staff organisation, "CD" (SOE) was not, despite the fact that SOE's role was operational. They therefore recommended that the Chiefs of Staff once again investigate SOE-SIS coordination. On the substantive issue of priorities however, they supported SIS's claim for priority in France.[35]

This episode illustrates two main points. First, by mid-1943 there was still no satisfactory coordination of SOE-SIS activities, which was badly needed because of obvious rivalries and conflicts of interest; and, second, SIS, through its seniority, position, and authority within the Chiefs of Staff machinery and elsewhere in Whitehall, enjoyed considerable political advantage over SOE in bureaucratic disputes. And this was Menzies' real forte. He may have been, as Trevor-Roper has claimed, "a man of very narrow horizons,"[36] but, as Philby observed, "his real strength lay in a sensitive perception of the currents of Whitehall politics, in an ability to feel his way thourhg the mazy corridors of power."[37] Alliances between the Air Ministry (particularly Bomber Command) and SIS to thwart SOE were not uncommon, and given Bomber Command's political weight this was no mean advantage.

The workings of this "alliance" can be glimpsed again, for example, early in 1944, following a characteristically impetuous intervention by Churchill in January to give greatly increased support to the French *maquis*. Almost immediately, Sir Claude Dansey of SIS recorded and circulated a conversation with Dewavrin in which the latter was alleged to have doubted "whether as many as 3000 men could be found anywhere in France who would be prepared to act together in a guerrilla offensive at invasion time." This produced, in M.R.D. Foot's words, "one of the strongest of many efforts made by other secret services to get the air effort for SOE cut down, on the grounds that it intensified Gestapo activity and so endangered all clandestine agents.[38] The SIS campaign also extended to PWE activities in France,[39] and reached such an intensity that Churchill complained to Ismay that "someone has been stirring him ["C"] up and making mischief." The finger of suspicion pointed to Bomber Command, strongly opposed to the diversion of aircraft.

The culmination of SIS-SOE rivalry came in December 1943, marking the climax to a train of events which had begun with the July-August crisis over aircraft. In September, SOE's handling of the Greek resistance had so alienated the Foreign Office and its representatives and allies in Cairo that it led to a special meeting of Ministers, chaired personally by Churchill on 30 September, which reaffirmed that SOE "would preserve its integrity under the Minister of Economic Warfare."[41] Hambro was forced to resign and was replaced by Gubbins, and shortly afterwards COSSAC assumed operational control of SOE activities in North-West Europe. But the Foreign Office was now deeply hostile (it had never regarded SOE with benevolence), and in December the opportunity came for it to join hands with SIS and elements of the regular military establishment to rid themselves of it. These forces were encouraged to act, moveover, by Churchill's prolonged absence from London, first for the Teheran Conference and then because of his serious illness and convalescence in North Africa.

On 1 December 1943 Air Marshall Bottomley informed the Chiefs of Staff Committee that the Commander-in-Chief, Bomber Command (Harris) had suspended all SOE flights to occupied Europe. The Air Ministry had received information that the entire SOE organization in Holland had been penetrated by the Germans, and this probably accounted for an abnormally high ratio of losses suffered by Bomber Command in carrying out SOE operations in that area. Even more serious, Bottomley told the Committee that "SIS had warned SOE, on more than one occasion that there was a danger of what had in fact occurred, but SOE had refused to accept their advice." The fear that this situation might be duplicated throughout Europe had led to Bomber Command's decision to suspend all SOE operations.[42]

Here was a familiar pattern: SIS information indicating that SOE networks in Europe were insecure produced without warning by the Air Ministry at a Chiefs of Staff Committee meeting; in this case, however, not to seek a reduction or prevent an increase in SOE flights, but to justify a suspension already ordered unilaterally by Bomber Command. As before, the situation was referred both to the Defence and Joint Intelligence Committees, and a full-scale enquiry into SOE operations in Europe followed. This took several weeks, and it was not until mid-January 1944 that the Defense Committee met under Attlee's chairmanship to consider the results. Three of the JIC conclusions concern us here. First, it recommended major changes in the organisation of SOE's relationship to the higher command; these were deferred for consideration until Churchill's return, when he once again reaffirmed SOE's independent existence under its own Minister. Second, it revealed that the SOE organisation in Holland had been penetrated and operated by the Germand for over a year, but that the situation elsewhere in Europe, with minor exceptions which had been or were being corrected, was satisfactory. And third, it became clear that SIS's role in the Dutch affair was little better than that of SOE; as the Secretary of State for Air himself admitted, "there had been a great deal of wisdom after the event in this matter." SIS has been consulted and in the summer of 1943 had been unable to produce any evidence of penetration of the SOE groups in Holland."[43] That the affair had provided the opportunity for considerable mudslinging between SIS and SOE is apparent from the reactions of the Chiefs of Staff to the Report. Sir Alan Brook, the CIGS (Chief of the Imperial General Staff), considered the situation so serious that he thought that "SIS and SOE should be brought under one ministerial head." Ideally, both should be placed under the Minister of Defense, but if SIS could not be removed from Foreign Office control, then "he would rather see SOE come under the Foreign Secretary than that the present situation should continue." He persuaded his colleagues on the Committee to support this view, and they recommended as such to the Prime Minister.[44] As we have noted, however, Churchill reaffirmed the status quo when he returned to London at the end of the month, and, temporarily consumed by a passion for the *maquis*, dismissed SIS rumblings of discontent as the result of Bomber Command machinations.

SIS's concern about SOE security, however justified it may have been, was not exclusively focused on the dangers of German penetration. By the second half of 1943, SIS's traditional and historic role as a clandestine anti-Bolshevik service was beginning to reassert itself, in tune with the increasing place which concerns about the postwar world played in the minds of those responsible for postwar strategy and diplomacy. Shortly before the incident over German penetration of SOE networks in Europe, fears that there had been a successful Communist penetration of SOE in London were raised when Captain B. L. Uren, "an army officer on special duty," was sentenced on 21 October 1943 to seven years imprisonment on a charge of having passed secret information to a high-ranking member of the Communist Party of Great Britain, Douglas Springhall.[45] The Uren case led to a purge of known Communists from SOE, and Churchill told Duff Cooper early in April 1944 that "we are weeding out remoreselessly every single known Communist from all our secret organizations."[46] As this remark indicated, SIS was also a target of the purge. It was within a few weeks of Uren's court-martial that an anti-Communist counter espionage section was established within SIS. From then on SIS was to focus increasingly on the post war intelligence tasks of the country, where the first priority target was the Soviet Union. The irony was that the man appointed to head this new section of SIS was Kim Philby, who had been a Soviet agent since the mid-1930s. Lord Selborne's defence of SOE against its critics within SIS in January 1944—"the danger of penetration is inherent in the work of a secret organisation, as I think SIS will confirm"[47]—was unwittingly but painfully accurate.

As the war drew to its close and as the Ministry of Economic Warfare prepared for its dissolution with the advent of peace, the question of the future of special operations arose. Would they continue, and, if so, under whose auspices and control? Would SOE continue in the post war era as an independent organisation, or would special operations revert to SIS? What role would the armed forces and the guardian of Britain's diplomatic interests play?

Anthony Eden and the Foreign Office had held strong views about the need to control special operations during the war, and on these post war issues Eden made his views clear to the Prime Minister late in November 1944. There would be a need in peacetime for a covert organisation to further the policy of the Government. Therefore, he told Churchill, "I should be sorry to see the abandonment of all machinery for 'special operations' even when the war is over." The wartime experience had convinced him, however, that special operations and the SIS should be under the same controlling head. "Nothing but chaos," he said, "can ensue if we try to have two secret organisations not under the same control working in foreign countries in peace time."[48] SIS in turn would be closely guided in its activities, as was traditional, by the Foreign Office. Eden therefore proposed that on Selborne's resignation as Minister of Economic Warfare on the conclusion of European hostilities, he (Eden) would assume responsibility for SOE, thereby asserting Foreign Office control of policy in Allied and neutral countries—although this would not interfere with directives issued to SOE by the Chiefs of Staff and Commanders-in-Chief in enemy or enemy-occupied countries. Despite Cadogan's reservations about Eden's proposals ("I am concerned that the F.O. should take over all these fantastic things. We aren't a Department Store"[49]), they formed the basis of the arrangements effected for the immediate post war period. But at the insistence of the Chiefs of Staff, who hoped that ultimately SOE and SIS would come under the control of the Minister of Defence, the measure was adopted as a short-term one, and SOE was to continue to work under its own Head (Gubbins), not under "C" (Menzies).[50] This last arrangement, the continuing organisational separation of SOE and SIS established in July 1940, remained undisturbed until after the war with Japan was concluded. On 14 August 1945, the day the Japanese Emperor decided upon Japan's surrender, the British Chiefs of Staff approved the recommendations of an ad hoc Committee on the Future of SOE, which called for its amalgamation with SIS under a common head, although there would be separate special operations and secret intelligence branches.[51] The longer term question of the control of SIS by Foreign Office or Ministry of Defence was once again deferred, this time until the completion of a major review of the British intelligence services by Sir Findlater Stewart. The Defence Committee accepted these recommendations on 31 August 1945,[52] and with subsequent arrangements with Menzies for the merging of the relevant sections of SOE with SIS early in 1946,[53] SOE's independent existence came to an end when it was finally wound up on 30 June 1946.[54]

If SOE was dead, special operations were not, and the final ironic act in the troubled war time relationship between SOE and SIS had yet to be played out. The special operations branch of SIS was soon involved in planning both for resistance inside Turkey in the event of Russian occupation, and in subversive activities inside both the Soviet Union (particularly the Ukraine) and in Communist Eastern Europe.[55] In an operation against Albania conducted jointly with the CIA from 1946 to 1952, designed to create an anti-communist resistance movement out of the defeated forces of the wartime civil war, SIS met with disaster. Practically all the agents infiltrated into the country were immediately arrested by the Communists and executed, and the operation was a total failure. Run in detail by ex-SOE personnel and reflecting the same assumptions which had underlined SOE's raison d'etre, the operation nonetheless was under the control of the professional SIS. It came under Philby's ultimate control and he betrayed its secrets to the Russians.[56] Thus a tragic symmetry of sorts between SOE and SIS was finally achieved: the German penetration of SOE which provided SIS with the ammunition for its strongest offensive against SOE was in turn matched by the Soviet penetration of SIS, and in both cases they resulted in the destruction of networks and the deaths of agents involved in attempts to foster resistance movements in Europe.

Three main conclusions may be drawn from this necessarily circumstantial account of SOE-SIS wartime relations. In the first place, their relationship was characterized by deep hostility on the part of SIS towards SOE. Although the evidence does not exist to reach more than tentative conclusions, it is difficult to imagine that this did not have operationally negative consequences for the war effort, although the extent and nature of these can only be guessed at. Second, attempts were made to improve these relations, but on each occasion the decision was reached that the existence of an independent SOE under its own Minister was necessary, despite the reservations of the Chiefs of Staff, the Foreign Office, and SIS. On at least two occasions, in September 1943 and again in January 1944, Churchill's personal intervention appears to have been decisive. Churchill, indeed, took the resigned view that "the warfare between SOE and SIS . . . is a lamentable but perhaps inevitable feature of our affairs."[57] Third and finally, the analysis of SOE-SIS relations emphasises the limits imposed upon Britain's policies towards European resistance movements both by the scarcity of its own resources and by operational priorities. It was not merely that SOE was a new organisation starved of aircraft which were central to its operations, as has often been argued, but that within the context of British strategy towards Europe, intelligence was given in most places and for most of the time a higher priority than resistance by those responsible for defining Britain's strategic policy. Thus this study confirms indirectly what the recent revelations about Special Intelligence ("Ultra") have indicated: the extraordinarily high value placed upon secret intelligence by the makers of British strategy in World War II.

Notes

1. Research on which this article is based was assisted by a Canada Council Leave Fellowship and Research Grant.
2. Quoted in Kim Philby, *My Silent War* (London 1968), 27; the words are those of an agent on transferring from SOE to SIS.
3. Donald McLachlan, "Intelligence, the Common Denominator", in ed., Michael Elliott-Bateman, *The Fourth Dimension of Warfare*, Volume 1, *Intelligence, Subversion, Resistance*, (Manchester 1970), 53.
4. The archives of both SIS and SOE remain closed. There are however numerous references to SIS and SOE activities in the released series of wartime papers in the Public Record Office, particularly in the Chiefs of Staff and Foreign Office Papers. A certain amount can be inferred from such sources, and it is on these, and from other written and oral sources, that this account is based.
6. See F. W. Winterbotham, *The Ultra Secret* (London 1974); Antony Cave-Brown, *Bodyguard of Lies* (London 1975); William Stevenson, *A Man Called Intrepid* (New York 1976). For the struggle to gain credit for *Ultra*, see H. Trevor-Roper's reviews of the last two books in the *New York Review of Books*, 19 Feb., 13 May 1976.
7. Christopher Andrew, "The British Secret Service and Anglo-Soviet Relations in the 1920s," *Historical Journal*, vol. 20, no. 3, 1977. For the Leeds seminar, see *The Times* (London) 4 and 6 February 1978.
8. For much of this background, see Bruce Page, David Leitch, Phillip Knightley, *Philby, The Spy Who Betrayed a Generation* (London 1968), *passim*; Patrick Seale and Maureen McConville, *Philby, the Long Road to Moscow* (London 1973) *passim*; Richard Deacon, *A History of the British Secret Service* (London 1969), 280-281.
9. M.R.D. Foot, *SOE in France* (London 1966), 1-3.
10. Philby, *op.cit.*, 4.
11. *Ibid*, 5.
12. For the "British Resistance Movement", see David Lampe, *The Last Ditch* (London 1968), and Duff Hart-Davis, *Peter Fleming, a Biography* (London 1974), 233-237. One participant described the operation as "pure Boy's Own Paper stuff."
13. M.I.(R) of the War Office had been created about the same time as Section D and was responsible for the planning of paramilitary activities. "EH" (Elec-House) was headed by Sir Campbell Stuart and concentrated on propaganda.
14. Hugh Dalton, *The Fateful Years* (London 1957), 366. See also the present author's articles on the origins of SOE: "The Detonator Concept: SOE, European Resistance, and the Fall of France", *Journal of Contemporary History*, Vol. 10, April 1975, and "Britain looks at Europe, 1940: Some Origins of SOE", *Canadian Journal of History*, Vol. 10, 1975.
15. And, it should be added, from the War Office, which resented the loss of control over MI(R). Both the War Office and SIS thereby conceived considerable hostility towards Dalton's organization.
16. E. H. Cookridge, *The Third Man* (London 1968), p. 91, and *Inside SOE* (London 1965), 46.
17. Philby, *My Silent War*, 12. See also Lord Gladwyn (Gladwyn Jebb), *Memoirs* (London 1972).
18. Author's interview with Major Grand, 12 May 1974, and *Dalton Papers* (Miscellaneous 1940) in the British Library of Political and Economic Science, London School of Economics and Political Science.
19. Sweet-Escott, *op. cit.*, *passim*. SOE was allowed, however, to use its own ciphers.
20. See the record of a meeting between Dalton and Mountbatten on 9 January 1942 in *The*

Dalton Papers. For the situation in November 1940 see *FO* 898/9. The late Brigadier Sir Richard Gambier Parry KCMG (1894-1965) became the post war Director of Foreign Service communications. Another area where SOE depended heavily on SIS was in the forging of false identity papers for use by agents abroad. Here, too, it was not until after some time that SOE established its own organization to handle such matters.

21. "All intelligence received by SOE from its own sources is, by charter, passed direct to SIS" Lord Selborne (Minister of Economic Warfare) on 11 January 1944 in *CAB* 69/6 [Public Records Office, London].
22. Ewen Montagu, *Beyond Top Secret Ultra* (New York 1978), 94-95.
23. See Richard Clogg and Phyllis Auty (eds.), *British Policy Towards Resistance in Greece and Yugoslavia* (London 1973), 209-214.
24. M.R.D. Foot mentions a case in 1941 in which Dalton claimed that SOE had wrecked a train in Hungary, when they had no agents there: "M16 aware of this, took care he was found out." See *Resistance: An Analysis of European Resistance to Nazism, 1940-1945* (London 1976), 201.
25. Foot, *SOE in France*, 66-7.
26. *Ibid.*, 104. Its first transmitting station was at Grendon in Buckinghamshire.
27. Sweet-Escott, *op. cit.*, 123-124.
28. JIC(42)156(0)(Final) of 29 April 1942, in *CAB* 84/85.
29. JP(42)470(S) of 2 May 1942, in *CAB* 84/85; and COS (42)15]th in *CAB* 70/20.
30. "SOE Collaboration in Operations on the Continent," COS(42)133(0) of 12 May 1942 in *CAB* 80/62. The "SOE Directive on Subversion in North Africa" of the same date was more explicit, instructing SOE that "You will maintain the closest touch with SIS at all times to ensure that your respective activities do not clash, and that the latter's sources of information are not imperilled." See COS(42)134(0) in *CAB* 80/62.
31. "Special Operations Directive for 1943", 20 March 1943, COS(43)142(0), *CAB* 80/68.
32. For example, whereas 108 SIS agents were operative in Belgium in the period 1940-1945, there were only 77 SOE agents during the same period. See G. Lovinfosse, "Belgium and Britain", *Great Britain and European Resistance* (St. Anthony's 1962).
33. COS(43)404(0) of 25 July 1943 in *CAB* 80/72.
34. COS(43)178th(0) in *CAB* 79/63.
35. JIC(43)325(0) of 1 August 1943, entitled "SOE Activities in France", in *CAB* 79/63. The entire question was then discussed at a Defence Committee meeting on 2 August, but the record of this meeting remains closed.
36. *New York Review of Books*, 19 February 1976.
37. Philby, *op. cit.*, 82. SIS was also represented, for example, on the XX ("Twenty") Committee and on the "W" Board which dealt with double agents. SOE was not, except on an ad hoc basis. See Sir Findlater Stewart to Sir Frank Nelson, March 1942, quoted by J.C. Masterman, *The Double-Cross System in the War of 1939 to 1945* (Yale 1972), 62.
38. Foot, *SOE in France*, 354-355.
39. See COS(44)172(0) in *CAB* 80/80.
40. Churchill to Ismay 10 February 1944 in *PREM* 3 185/1. SIS was undoubtedly assisted by the less than benevolent observations on SOE activities passed on by Sir Desmond Morton, Churchill's personal assistant, who from the early days of the war had been charged with special responsibilities for French affairs. Morton had a long-standing relationship with SIS, and rarely failed to draw Churchill's attention to SOE failures. See the *PREM* 3 185/1 file.
41. COS(43)594(0) of 30 September 1943 in *CAB* 80/75, and COS(43)618(0) of 11 October 1943, also in *CAB* 80/75.

42. COS(43)293rd of 1 December 1943 in *CAB* 79/88.
43. The JIC report remains classified; but it was discussed by the Defence Committee on 14 January 1944, the minutes of which are available in DO(44)2nd(Final)(0) in *CAB* 69/6; and Selborne's reply to the JIC report [DO(44)2 of 11 January] is attached to the minutes.
44. COS(44)1st(0) of 3 January 1944, *CAB* 79/89.
45. See *The Times* (London) 8 November 1943. Uren was court-martialed on 21 October 1943, and it is clear that he worked for SOE. Springhall had been London district organiser for the CPGB in the late 1930s, was British representative at Comintern HQ in 1939, and in 1943 had become national organiser of the party and a member of its Politburo. He had been responsible for recruitment to Soviet espionage networks throughout the 1930s, was convicted on 28 July 1943 to 7 years period servitude for obtaining secrets from an Air Ministry employee, and "it later transpired that he had also been receiving information from an army officer who was engaged in secret work." Henry Pelling, *The British Communist Party, a Historical Profile* (London 1958). The army officer was Uren. See also E.H. Cookridge, *The Third Man* (London 1968), 19.
46. Churchill to Duff Cooper, 6 April 1944. SOE 44/17, in the *Avon Papers*. [PRO], *FO* 954/24.
47. "SOE Operations in Europe," memorandum by the Minister of Economic Warfare 11 January 1944. DO(44)2 in *CAB* 69/6.
48. Eden to Churchill, 23 November 1944, COS(44)381st(0), in *CAB* 79/83.
49. *Diaries of Sir Alexander Cadogan, 1938-1945*, ed. David Dilks (London 1976), 683.
50. COS(4)381st(0) of 27 November 1944 in *CAB* 79/83. Subsequent comments by Eden on this arrangement remain classified. See COS(44)389th(0), *loc. cit.*
51. COS(45)198th of 14 August 1945 in *CAB* 79/37.
52. DO(45)4th of 31 August 1945 in *CAB* 69/71. The memorandum involved remains closed until 1996.
53. COS(45)389th of 27 December 1945, *CAB* 79/42. The ad hoc Committee was chaired by Cavendish-Bentinck, the F.O. representative and Chairman of the Joint Intelligence Committee.
54. COS(46)58th of 11 April 1946, *CAB* 79/47.
55. Philby, *op. cit.*, 103, 119-121.
56. Bruce Page et al, *op. cit.*, 194; Anthony Cave-Brown, *Bodyguard of Lies* (New York 1975), 807.
57. Churchill to Ismay, 10 February 1944, in *PREM3* 185/1.

INTELLIGENCE

Psychological Warfare and Newspaper Control in British-Occupied Germany: A Personal Account

Frank Eyck
University of Calgary

Some autobiographical detail is necessary to understand how I became involved in psychological warfare and in the British army of occupation in Germany. I was born in Berlin in 1921 as the son of Dr. Erich Eyck, then a lawyer, journalist and politician, later a historian, and of his wife Hedwig, nee Kosterlitz. Both my parents were Jewish and under the Nazi regime our family was persecuted. Fortunately my parents, my sisters, and I, as well as many members of the closer family, were able to emigrate. I left the Franzosische Gymnasium (College Francais), where tuition was in French, in 1936, to continue my education at St. Paul's School in London. By the outbreak of the war I was able to speak not only my native German, but also English, and I retained something of my French. I joined the British Army in 1940 and as a non-Britisher was first of all subjected to a fitness course consisting of two years in the Pioneer Corps loading and unloading railway trucks, concreting roads, putting up Nissen huts, etc.

There was, however, an increasing agitation against the waste of man-power constituted by lawyers, medical doctors, as well as younger men with a good schooling like myself being assigned to navvies' tasks. Most of us wanted to be more useful, and we received splendid support from outside the army. Somebody gave my name to one of the many helpful spirits active on our behalf, Mrs. Dorothy Buxton, the widow of the Liberal and later Labour member of Parliament Charles Roden Buxton. Mrs. Buxton quite systematically built up a dossier by writing to non-British soldiers in the Pioneer Corps, like myself, and from 1942 onwards it became easier to obtain other employment in the Services. In the fall of 1942, I was attached to the Army Educational Corps, and a rather sudden transition found me lecturing on the 'British Way and Purpose.' For about a year and a quarter I motorcycled all over one English county, Northamptonshire in the Midlands, giving talks to British army units. In the weeks before D-Day I was recalled to the Pioneer Corps and subjected once more to the delights of the parade ground. Given the opportunity to apply for other postings, I survived the Alpine Rope Climb Race, but nearly broke my neck in a fall. The army authorities wisely decided to earmark me for specialist employment rather than attempt to make an infantry officer out of me. Just after D-Day I was interviewed in London for an advanced intelligence unit which was to operate on motor-bikes, but which was not actually formed, perhaps a victim to the rather rapid Allied advance in North-Western Europe. However, I received a posting to 21 Army Group Headquarters. 21 Army Group, the British army group, was commanded by Montgomery. Headquarters consisted of three parts. 'Advanced' was already in Normandy with Montgomery. I joined 'Main' in the Portsmouth area. There was also 'Rear.' Main H.Q. moved to Normandy on D+60 early in August 1944.

At 21 Army Group Headquarters, I was posted to Publicity and Psychological Warfare Branch (P & PW). The Psychological Warfare side of the branch was engaged in a number of propaganda activities, to several of which I was allotted in turn during the following months. I began with monitoring. For this I had been tested at the BBC in London during the summer, while flying bombs descended on the city. The examiner was merciful with me. I was asked what the German for 'monitoring' was, and only found out later that it was 'abhoren'. I was tested in both German and French, the latter by that time having got rather rusty. In Normandy I was one of a team listening in initially to occupied France, where French collaborators broadcast anti-Allied propaganda in French on stations under German control. We made notes in our van, but for anything very special we used recording facilities. I was very relieved when later in August 1944 I heard English voices from Paris, of war correspondents who had somehow got to the microphone before the French capital was officially liberated. My job became easier when, with France in Allied hands, I was switched to monitoring German radio stations in German. I often heard Joseph Goebbels, the Nazi minister of propaganda and enlightenment, who to my surprise sounded very reasonable and even God-fearing. According to him the Germans were being treated badly by the other nations, who completely misunderstood them. If one had not known about reality, it would have been very convincing.

Our reports were carefully analysed by the psychological warfare officers in our branch at HQ and anything particularly interesting was passed on for further intelligence assessment. The broadcasts provided clues to enemy morale and told the experts something of the way the Nazis visualized their situation. From what I saw and heard the standard reached by intelligence officers at various levels was very high. I only hope that young amateurs like me and most of my colleagues provided the raw material for which the analysts were looking.

In August 1944, during the final stages of the Falaise gap battle, I was given a fascinating assignment. I was to come into contact for the first time with German prisoners of war. Two of us were briefed by an intelligence officer and taken by him to a large field where many German prisoners of war were behind barbed wire. We were told in strict secrecy about a radio station masquerading as German but run by the Allies. The French port of Calais was in German hands and the station, which addressed itself to German soldiers in France, was called "Soldatensender Calais". It was used to try to undermine the morale of the German soldiers. While ostensibly it looked at things from the German point of view, now and then items were introduced to create uncertainty and fear among the German troops. To authenticate the news given on the station, it was important to include some details about the designation of their units, their commanding officers, etc. Our job when interrogating prisoners of war was to obtain any information we could on the units to which they belonged and on other units in their sector of the front. If one could dig up anything disreputable about members of formations still in the front line, this might be slipped into the broadcasts, seemingly almost by oversight. Presumably military intelligence also supplied information for the station.

It was very curious for a German refugee, on the extermination list and furnished with a "nom de guerre" to avoid detection in case of capture, to find himself face to face with the German soldiery. Still remembering the demeanor of the Hitler Youth before my emigration and having read plenty about the fanaticism, particularly of the younger troops, I was prepared for great difficulties in my interviews. To my surprise the German prisoners of war were quite ready to talk. So far as they were concerned, the war was over and they wanted to make a fresh start. In general they did not then identify themselves with the Nazi regime, even if, presumably, most of them had done so earlier. If there was any indoctrination, it was quickly gone once the iron grip of Nazi propaganda had been removed. If the prisoners were surprised by a soldier in British uniform suddenly addressing them in fluent German, they did not show it. I do not know how successful we were in obtaining information, because as a soldier I was doing a limited job and did not see the end result. I do not know how much of the intelligence I provided was used by Soldatensender Calais. I hoped that I asked the right questions and that the prisoners, if they knew, gave me correct answers.

In discussions and interviews later, the question was often raised whether the kind of deceit carried on by Soldatensender Calais could be justified. While in the case of "white propaganda" those at the receiving end could identify the source, "black propaganda" tricked them carefully as to the originator. Quite recently a critic of the British Broadcasting Corporation in England attributed what she regarded as the decay of the B.B.C. to the activities of a former senior official who, she alleged, had participated in "black propaganda" during the Second World War (which was, however, denied). There are probably at least two aspects to the problem. First, was the use of this weapon against the Nazis defensible? My answer is an unhesitating "yes", and I personally never had any qualms when assisting in a small way in promoting "black propaganda". Second, there is the effect that participating in deceit has on the individual. I should have thought that in this respect most people kept their military and civilian activities in watertight compartments.

The Psychological Warfare section of P & PW Branch of 21 Army Group also ran, through Second Army Headquarters, a number of forward units. In about October 1944, after we had got to Brussels, I was assigned to one of these. Amplifier units were used close to enemy formations, attempting to talk to them by means of loudspeakers. They were fired on by German troops. Leaflet units advised on the drawing up of leaflets which could be shot over to the German lines by artillery. They were probably the smallest units in the whole of the British army, consisting of a German-speaking corporal, a driver and a small truck. The amplifier units were slightly larger. Normally an amplifier and a leaflet unit formed a pair and all the German-speaking personnel participated in the loudspeaker broadcasts when they took place. Some successes in speeding up surrenders were obtained, unfortunately at the cost of occasional casualties. The pair of units to which I belonged spent some time near the front line in the Roermond region, was withdrawn at about the time of the German Ardennes offensive in December 1944, then moved back to 21 Army Group HQ in Brussels. I was never used, like many soldiers spending my time merely waiting. Before the final push into Germany I was assigned to other duties.

Political warfare, including psychological warfare or propaganda, is subordinate to military warfare. Propaganda cannot in itself win wars; in a well-ordered campaign it must be fitted into military strategy. The military and the psychological arms of warfare, in my low-level experience, worked well together. Was the expenditure of manpower and resources on political warfare, including propaganda, worth while? To take one extreme answer: it would have been unthinkable, in view of the importance of Nazi propaganda, to have ignored it altogether and leave the field to it. Some political warfare was essential. The effort spent on the "white propaganda" of the BBC in London was well worth while. It became clear from conversations with Germans after the end of the fighting that the BBC had established for itself a standard of accuracy and reliability which proved of great importance for the future.

The question can thus be narrowed down to certain elements of propaganda. There was the dropping of leaflets over enemy-held territory from aircraft. Probably some of the effort on this was wasted, but one type of leaflet scored considerable successes, the *Passierschein*, the pass issued by the allied armies which made it easier for Germans to surrender. While the forward leaflet units could in theory move into action more quickly than the political warfare organisation in England, it was found more convenient and perhaps wiser from a policy point of view to entrust operations mainly to those who were thoroughly experienced, even if remote from the front line. The amplifier units could operate best if the enemy was cut off, and by speeding up a surrender could save lives and time. There *might* have been situations in which the forward units could have been used more intensively. The manpower and resources had to be there in case of need. A spin-off effect was that psychological warfare in all its ramifications, from the collection to the diffusion of information, helped to train a body of specialists likely to be useful in future contingencies, including an army of occupation.

That leaves the controversial matter of "black propaganda". It would be hard to prove—or to disprove—the effectiveness of this particular method of warfare. Unless it was too expensive or wasteful of manpower and resources, it was worth while and may have been of some benefit.

In February 1945 I was recalled to England to be trained for the army of occupation in Germany. The Allied authorities were hopeful the war would end during the campaign of 1945 and felt that to secure an efficient occupation a small number of key personnel could be spared from the armies in the field. Just as my experience in army education may have been useful in preparing me for psychological warfare, which has been regarded by some as being close to education, so psychological warfare had certain affinities with control of the media in an occupation regime. The latter two were both concerned with the effects of Nazi indoctrination in various ways. Psychological warfare had to assume at least some Nazi indoctrination in most of the recipients, occasionally trying to exploit it. The army of occupation had to seek to remove it, possibly by indoctrinating something else. In any case, experience in the wartime activity was of some use in the peacetime pursuit. An information control branch was founded, recruited from various sources, including psychological warfare personnel. Training took place in a course given near London, where a wide variety of exceptionally well qualified specialists lectured on German history, politics, psychology, sociology, etc. The course was among the most intellectually stimulating experiences I have ever had. A whole new world was opened to me. In April 1945 we were transported back to the continent and soon crossed the border into Germany.

It felt strange to reenter Germany with a victorious invading army after fleeing from persecution a few years earlier. The war was not yet over and we still could not be sure of being safe. There were reports that there were pockets of Nazi resistance in the areas to which we were moving. We had strict orders to remain armed. Certainly there was no occasion to be joyful, even if there was a sense of relief that the German army was losing its offensive capability and the war appeared to be nearing its end. The sights we saw on our way to Westphalia were sombre. The destruction of the cities was the most effective commentary on Nazi propaganda. How did members of my unit—many of whom were also German refugees—and I feel about returning in these circumstances? How did we view the German people? To what extent did we regard the German people as responsible for the persecution of the Jews and of other enemies of the Nazis? Unlike civilian refugees, we had to take the plunge. We were under orders and were confronted with a situation to which there are few parallels in history. Many civilian refugees hesitated for long about even visiting Germany and some never did, while others returned to settle as soon as it was feasible.

When I think back now, particularly with hindsight, I certainly did not appreciate all the complexities of the situation in which the Germans who remained under Hitler found themselves. I personally knew a number of anti-Nazis. But so far as the German people as a whole was concerned, I assumed that they had supported the Nazis, though some more willingly than others. To outsiders, the Nazi dictatorship presented an image of enthusiastic national unity and at least acceptance of the measures of the regime by all except for a small minority. I realize now that I saw things in too absolute terms before I was once more in actual contact with German civilians. The whole problem in April 1945 resolved itself for me into the question of whether Germans outside the ranks of the Nazi hierarchy knew about the dreadful treatment of Jews and other persecuted groups. I was reluctant to believe that one could live in Nazi Germany without knowing at least something about these things. And if people knew, what did they do about it? Did they in their hearts condemn these happenings? Perhaps my attitude before the end of the war can be summed up as very critical of the German people, averse to accepting any excuses, full of reserve for a nation which had excluded me and my kin, natives, from the rights of citizenship. But I was free from the kind of hostility which would have made me determined to inflict harm on the German people. The whole situation was far too serious for vindictiveness. For reasons of faith I had been baptised, as a soldier, on the day after D-Day, soon after receiving orders for overseas mobilisation. But the German Jews were my people, the group into which I had been born, and I felt with them and for them. By April 1945 enough was known about the persecution of the Jews to be certain that annihilation had been carried out on a vast scale. Could I, fortunate enough to escape, forget what had been done to my people? How could I ever feel at ease in Germany? The next year supplied many of the answers.

The Western authorities took Nazi propaganda seriously and were determined to bring the media under Allied control. In the occupied areas no German newspapers or radio stations were allowed, a policy which continued for a time after the end of the fighting in Europe. The Allied armies themselves manned the media. Within my particular information control unit, I was one of those allocated to the newspaper side. In Westphalia, in a little place southwest of Osnabruck called Lengerich, I belonged to a small staff that wrote and published a newspaper in German for the civilian population. When the German armies surrendered early in May 1945 I was among members of the unit moved up to Hamburg. We concentrated on Hamburg and Schleswig-Holstein, but for a time also looked after some regions further to the East which were later surrendered to the Russians.

When I was being trained for the army of occupation I had hoped I would be sent to Hamburg. It had been one of the last places I had visited prior to leaving Germany before the war. It was a city with which my family had many connections, where one branch had lived and we had several friends. The port of Hamburg had long maintained close relations with England. In the prewar period, Nazi persecution had not been quite so harsh there as in most other parts of Germany.

Destruction had been severe. Allied bombing raids had almost wiped out parts of the city. We took over the printing press of one of the newspapers and quickly produced our own newspaper. We gave the main news, which was largely a record of Allied measures, the setting up of the Allied Control Council, and so forth. The newspaper was avidly read by the civilian population, which was short of hard news. Our work was not made easier by the non-fraternisation order issued by the British army authorities. We used some of the editorial staff of the newspaper we had taken over, and strictly speaking we were supposed to confine our conversation to what was necessary to carry out our duties. However, we were all so eager to find out how things had gone in Germany during our absence that our conversations hardly ceased. After a few weeks the non-fraternisation order, which had rapidly become a dead letter for the troops, was lifted. Most of us soon developed a circle of German friends, people with whom we discussed politics freely and in whose integrity we had confidence. On an individual basis the barriers were broken down very quickly. It was more difficult to form an impression of the attitude of the mass of the people, to make any generalisation about their political outlook and their views of the recent past. It was a great help gradually to resume contact with friends of one's pre-emigration days and to get some guidance from them about the intervening period. Many of these reunions were very moving, joyful for those who had survived, sad when thinking of members of families who had died before their time. One of our family friends had sheltered one of the conspirators of the 20th of July in her home and visited him in prison after he had surrendered, shortly before his execution. However much one had suffered, one felt very humble about such acts of courage.

Information control required confidence in the ability to turn the German people from the path they had followed during the Nazi regime. In our unit there was no place for those who felt that the Germans simply had to be written off, that "the best German was a dead German" (to take the extreme view). We were engaged in an educational job of a kind, summed up in the concept of reeducation, which may sound rather patronizing. We had to have the assurance of the educator that what we were putting forward was worth while and held out the hope of a better future. Both secular, humanistic methods and deep religious faith could provide the road to our goal. Both outlooks were represented in our unit.

It became clear during the weeks before the end of the fighting from information which some of us received that it would be very difficult to obtain agreement with the Russians on united occupation policy. Technically, however, Germany was occupied by the Allies and ruled by the Allied Control Council of the Western Powers and the Soviets. In our official German language newspapers in the British zone we were very careful, initially, not to criticise the Russians, particularly as we thought it would be bad for psychological reasons if the impact of the defeat on the Germans were lessened by the hope of quickly emerging from their isolation because of inter-Allied divisions. Our task was in a sense made more difficult and in another easier by the fissures in the Allied camp, which could not be kept from the Germans. When we publicized the main Nuremberg war crimes trial, no amount of journalistic skill could convince our readers that the governments which had appointed the judges were all strong supporters of democracy and of the right to fair trial, and that all war crimes and offences against humanity had been committed by the Germans. On the positive side the Germans in the Western zones on the whole regarded themselves as lucky to have escaped the Russians, and were therefore more amenable to our reasoning than they would otherwise have been. The question for the Germans was to what extent they needed "reeducation". Many Germans felt that they had been taught their lesson already and that there would be little trouble from them in the future. They had been cured of extreme nationalism and of admiration for dictatorship. They could get over their defeat more quickly, in a sense, than their fathers in 1918, because in 1945 a clearer distinction could be made between the regime and the people. The extent to which Germans made this distinction was a surprise to us. Those we met did not then feel any identity with the regime. While that attitude was satisfactory, if genuine, it left some questions unanswered. There were not many admissions of a change of heart. If the majority of Germans were not now Nazi, where had all the supporters of the regime gone? Whatever allowances one made, many of us regarded the change of attitude as rather too abrupt and somewhat undignified.

In our newspaper we argued the superiority of democracy over dictatorship. We inveighed against militarism, the latter so successfully that the obstacles to Germany's later entry into the Western defense network in the 1950s were increased. We enforced a strict separation of news and comment. This was an understandable reaction to the propaganda policies of the Nazi regime which had carefully avoided any such distinction. So far as possible our comment, in leaders and feature articles, was clearly marked off from the news, which was given in the most straightforward manner. The validity of the distinction between news and comment has come under strong challenge in recent years, but I still think that it makes sense and that we did a good job in promoting it in Germany after the war.

After a few months, information control withdrew in most places from running the newspapers and instead assisted in licensing them. Care was taken to have different points of view represented in a locality through the various political parties which were gradually permitted; in some places the Communist party received a license. Eventually party allegiance of newspapers decreased in importance; some licensed party political newspapers did not do at all well. The judgment shown in the choice of licenses varied. While the actual arrangements made for newspapers did not always last, British information control had a permanent influence in setting standards which contributed to improved newspapers. A cameraderie developed between information control personnel and German journalists and newspaper publishers, which often continued after demobilization. The pre-censorship and later the post-censorship of licensed newspapers before their eventual liberation was comparatively brief and did not, on the whole, create very great problems. Cooperation between the occupation authorities and the newspapers was close and in general smooth. It was a forerunner of collaboration in the European Economic Community between officials of various nations, for which it was in many ways a trial run.

I had an interlude from Hamburg when during the winter of 1945-46 I spent a few months in Flensburg, on the Danish border, helping to supervise the transition from direct to indirect control. There was a movement in the border region to join Denmark. The British authorities were determined to remain neutral over this attempt to revive the Schleswig-Holstein issue. During direct control no reference to this question of greatest interest to the border population was allowed.

I was fortunate after the tragic years of the Nazi regime to be able to assist in a small way in the work of reconstruction in the country which had threatened Europe and the world. By the time I was due for demobilisation from the army the most interesting and fruitful work had been done in the field with which I had been concerned. Some people stayed on, but that did not tempt me. Some older refugees settled in Germany, but very few young ones. My mother and father wanted to stay in Britain and it was a foregone conclusion for me that I would remain there after demobilization. I was released from the army in July 1946 and in October I joined the ranks of the undergraduates at Oxford.

Bibliographical Note

There is no official history of the Political Warfare Executive, which was responsible for psychological warfare. Charles Cruickshank recently published an interesting study of psychological warfare 1938-1945 under the title *The Fourth Arm* (London, 1977). Some relevant information on the topic may be found in Asa Briggs, *A History of Broadcasting in the United Kingdom*, vol. 3 (Oxford 1970); Sefton Delmer, *Black Boomerang* (New York, 1962); and David Lerner, *Psychological Warfare against Nazi Germany* (Boston, 1961). There are some useful articles in *The Journal of the Royal United Service Institution*, London: "Political Warfare" by Sir Robert Bruce Lockhart in vol. 95 (May 1950), 193-205, and "Psychological Warfare" by R. H. S. Crossman in vol. 97 (August 1952), 319-32. See also Michael Balfour, *Propaganda in War, 1939-1965* (London, 1979), and Charles Cruickshank, *Deception in World War II* (Oxford, 1979).

The British occupation of Germany has been widely treated. A good specialized book on the aspects covered in my article is Raymond Ebsworth, *Restoring Democracy in Germany* (London, 1960). See also Lord Barnetson, "The rebuilding of the German Press: 1945-1949" in

Journalism Studies Review, Cardiff, No. 3, June 1978, 7-11.

    This paper is dedicated to the memory of Sir John Masterman (1891-1977), who was Provost of Worcester College, Oxford, while I was an undergraduate, in thankfulness for the interest he took in me, even long after I had left the College. If he had lived, it would have been a privilege for me to have sent to the author of *The Double Cross System* (Yale University Press, 1972) this account of a modest contribution to some of the activities in which he excelled.

THE BOMBING WEAPON

The Royal Air Force and the Origins of Strategic Bombing.

Sydney Wise
Carleton University

No weapon of the twentieth century has done more to extend the war zone to civilian populations than the bombing aircraft. This development was foreseen well before the first bomber made its appearance; prophetic writers predicted its invention and foretold its use against the urban masses. Not until aviation technology had advanced somewhat did politicians and airmen take up seriously the idea of strategic bombing, as they came to see that the air arm possessed potentialities beyond purely tactical and subordinate roles. By 1916, during the First World War, responsible political and military authorities were to be found on both sides prepared to argue that the war could be won by striking through the air at the enemy's industrial might or at the national will by attacks upon the civilian population.

Much of the literature on the history of strategic bombing neglects the First World War. There has often been, for example, a tendency to attribute the first professional military expression of the idea to such interwar theorists of air power as the Italian military writer, General Giulio Douhet. The credit, if that is the word, for the first use of "terror" bombing against civilians has been variously given to the Italians in Ethiopia, or to the Germans in the Spanish Civil War, or to the Japanese in China. But even this development was foreshadowed, at least in planning, during the First World War.

The German High Command engaged in a form of strategic bombing from early 1915, and both the French air force and the British Royal Naval Air Service also undertook long-range bombing early. For the last twelve months of the war, a British formation waged a strategic bombing offensive against Germany. To a remarkable degree Germany and Britain thought through the problems of defense against the bomber, both in daylight and night operations, and by 1918 defensive systems comparable to those of the Second World War (with the obvious exception of radar) were in place in the two countries.

Though the experience of the First World War rarely is referred to in discussions of the history of strategic bombing, an examination of it reveals that weaknesses plainly visible in later bombing offensives were present from the beginning. With the obvious exception of the use of atomic weapons against Japan in 1945, strategic bombing has never achieved the aims of its advocates. Much controversy still surrounds the Allied bombing offensive against Germany during the Second World War, and a large professional, historical and polemical literature exists on the subject. In due course, a similar literature will doubtless be built up around the American bombing offensive in Southeast Asia. In the First World War as well, the hopes of the politicians and the objectives of the planners always outran the capacities of air crews, the availability of air strength and the state of aviation technology.

The First World War was a forcing house for a variety of weapon systems, but most notably for aircraft. Except for limited use in the Balkan Wars and by the Italians in North Africa, the airplane had never been employed in military operations prior to 1914. It had only been invented, after all, in 1903. But during the war every military possibility inherent in the aircraft, and since familiar in warfare, was seized upon and exploited. At the beginning of the war the airplane was almost exclusively an eye in the sky, reporting enemy dispositions and imposing the necessity of night movement and camouflage. Soon the linking of the airplane to the camera brought the detailed mapping (and constant re-mapping) of enemy defense systems and back-area communications. The airplane also became a pointer and director for field and heavy artillery, a role so vital in this war of sieges that by 1917 ninety percent of artillery observation in the Canadian Corps was performed by aircraft. By 1915 the airplane had become a gun platform, used first to strike at enemy aircraft and later at troops and other ground targets. By 1918, as at Amiens on 8-9 August, the airplane had become part of a complex combat team of infantry, artillery and tanks, tied together by wireless, that revolutionized tactics and restored movement to warfare. It was perhaps inevitable, then, that the airplane should also be used as an extension of artillery with a trajectory far exceeding the longest-ranging siege gun, an infinitely mobile platform from which fire and high explosive could be dropped.

Whereas most military uses of the airplane were adopted readily by armed forces, the employment of aircraft for bombing encountered resistance on a number of grounds. There was, of course, no opposition to tactical bombing; from the time of the battle of Loos in late 1915 such operations were always part of British battle plans. It was the use of aircraft to attack targets well removed from the actual theatre of ground operations that aroused doubts among professional soldiers and airmen and was to divide politicians, the press and sections of the British public.

Most of those who objected to the use of aircraft for bombing targets in built-up areas had moral reasons for doing so, because it was recognized from the beginning that civilian casualties would be unavoidable. Convention of long standing severely limited the ways in which acts of war could be visited upon civilian populations. Moreover, there was a healthy fear of retaliation in kind by the enemy. Though sharing these scruples and apprehensions, military men were much more likely to emphasize the technical limitations of the aircraft and the misuse of productive capacity that a diversion of effort to large scale bombing raids would bring about.

In a sense, the British public mind had long been prepared for the idea of destructive bombing raids upon great cities. In 1842 Tennyson had prophetically written of the "ghastly dew" that rained "from the nations' airy navies grappling in the central blue." Many popular writers, especially in the period immediately before the war, had taken up this theme, weaving terrifying visions of the death that would come from the skies. In 1908 H.G. Wells envisioned the destruction of New York and the collapse of the United States by a mere dozen bombing aircraft.[1] Many of those who advocated the employment of bombers to "win the war", politicians and airmen among them, echoed this fantastic literature when putting their case. To soldiers like Sir Douglas Haig and Sir William Robertson, the Chief of the Imperial General Staff, such notions were utterly absurd and founded upon a gross misunderstanding of the capabilities of the airplane.

As it happens, the soldiers were right and the politicians and the airmen were wrong. Yet by late 1917 the British Cabinet had decided upon a strategic bombing offensive against Germany, had created an independent air service, the Royal Air Force, in part to fulfill that purpose, and had allocated substantial resources towards the formation of a bomber fleet. The Cabinet took these decisions against the express advice of Sir Douglas Haig and Major General Hugh Trenchard, commanding the Royal Flying Corps in the field. The immediate explanation of the Cabinet's behaviour is to be found in the crisis induced by the German heavy bomber attacks of 1917, but momentum for these decisions had been building for at least two years.

The idea of using bombers to strike against Germany originated in the period of the first German airship attacks upon England. The rigid airship, in use commercially in Germany before the war, had not originally been though of as a bombing weapon. The German Naval Airship Division was established as a scouting and reconnaissance force for the High Seas Fleet, and the German Army possessed a similar unit. Shortly after the war began, however, arguments in favour of employing the airships to raid England began to be pushed at the highest level. The Deputy Chief of the German Naval Staff, for example, was persuaded that a few raids on the London docks would set off irrepressible panic and "render it doubtful that the war can be continued," an observation demonstrating that the "bomber mentality" and fantasies connected with it were by no means confined to the British. Wilhelm II had his doubts, but they were overcome by the wave of popular enthusiasm that greeted the news of the first limited raid on England in January, 1915. As a result, he issued the following order:

> 1. His Majesty the Kaiser has expressed great hopes that the air war against England will be carried out with the greatest energy.
> 2. His Majesty has designated as attack targets: war material of every kind, military establishments, barracks, and also oil and petroleum tanks and the London docks. No attack is to be made on the residential areas of London, or above all on royal palaces.[3]

It is notable, of course, that Wilhelm intended to bomb neither the Londoners nor his own relatives, but what is chiefly significant about his order is its gross optimism. Bomb-aiming was a crude art during the First World War; at this stage it was primitive. Moreover, the chief defense the airship had against ground fire and attack by hostile aircraft was altitude. An **awkward, slow-moving Zeppelin filled with highly inflammable hydrogen** was a huge and vulnerable target, as events were to prove. Airship captains, therefore, sought safety in altitude. The first operational airships had a ceiling of 11,000 feet, but bombing from this height was a hit-or-miss affair, especially when Zeppelin crews took to bombing through cloud layers. Despite the precision of their orders, their bombing of necessity was virtually indiscriminate, and heavy civilian casualties occurred in some raids.

Even though the casualties they caused and the property they destroyed were never great enough seriously to affect the British war effort, the Zeppelins were undoubtedly successful in achieving the limited strategic aims of the High Command. Despite the grandiose rhetoric which had accompanied their first use, the airships were never intended to accomplish more than a diversion of British air and ground strength from the Western Front. In other words, the German Naval Airship Division and the few Army airships which took part in raids upon England were used as auxiliaries to the operations of the German Army. In this sense they were almost classically successful. By the end of 1916 the Royal Flying Corps had committed eleven squadrons to home defense, and the War Office had built up a complex home defense organization, including air raid control zones, belts of searchlights and anti-aircraft guns, and a large ground observer corps and communications network. Thousands of men and much material had been kept from the battlefront in order to cope with the attacks of a handful of airships.

While the Germans had taken into account the psychological effect of airship raids upon England, they profoundly miscalculated its nature and consequences. There is no doubt that the initial impact was enormous. The appearance over England and Scotland of the huge and seemingly invulnerable Zeppelins, witnessed at one time or another by hundreds of thousands, was a gross and frightening demonstration of German power. There were many signs of fear and demoralization: factory workers downed tools and many stayed away from their shops for several days following a raid. There were a number of incidents in which uniformed members of the Royal Flying Corps were mobbed in the streets, apparently because the RFC had been ineffectual in its efforts to come to grips with the airships. Such manifestations of shock and diffuse anger never entirely disappeared during the German bombing offensives; there were a number of riots and other examples of mass panic. On the whole, however, the emotions released by German bombing tended in quite another direction: toward retaliation.[5]

The fate of the Naval Airship Division's L.19 illustrates the movement of public feeling. On 31 January 1916 nine airships raided the Midlands. L.19, probably because of engine trouble, came down in the North Sea near a British fishing trawler, the *King Stephen*. The captain of the trawler abandoned the crew of L.19 to death by drowning because he feared that if he took them aboard, he and his men would be overpowered. His action caused much controversy in Britain, but it was condoned by the Bishop of London because the Germans had bombed innocent civilians. So intense was the sense of moral outrage in Hull on the occasion of another raid in early March that it was fifty-four years before the Hull "Bobbers"—the stevedores who unload trawlers—permitted a German vessel to unload its catch at their port.[6]

Moral outrage led swiftly to the call for revenge, first in the press and at public meetings and then in Parliament, by such members as Pemberton-Billing and Joynson-Hicks. So powerful did this agitation become that the Convocation of Canterbury was moved to condemn it as immoral and barbarous. One Bishop dissented. The Bishop of Bangor, in dismissing the arguments of his brethren, argued that the new weaponry had converted all citizens into combatants. The best way to stop airship attacks, he suggested, would be to send a hundred aircraft to drop bombs "all over the rich business part of Frankfurt." For this view he was congratulated by C.G. Grey, editor of *The Aeroplane*, for "his intellectual honesty and freedom from cant."[7]

The failure of the defenses to repel airship attacks made the idea of retaliation irresistible to those in military and political authority. The first Zeppelin was not brought down over England until 31 March 1916, by anti-aircraft fire; the first success by a home defense aeroplace was on the night of 2-3 September, when Second Lieutenant William Leefe Robinson show down SL 11. Robinson's victory heralded the turn of the tide; his gun was armed with incendiary ammunition, lethal to the hydrogen-filled envelope of the airship, and from then on the German dirigibles fought a losing battle against steadily improving, better-armed aircraft and a more and more sophisticated air defense system.[8]

But before then the Zeppelins seemed invincible. Rear Admiral Vaughan Lee, Director of the Air Department in the Admiralty, argued in April that the best response to raiding was "an organised and systematic attack on the German at home," which he thought would not only cut down airship raids on England but also have "an immense moral effect on Germany itself." Both these assertions were accepted on faith by the Admiralty. Without reference to the War Office or the Cabinet, it set in motion the formation of a bombing wing of the Royal Naval Air Service. At the same time, through direct negotiation with French authorities, it secured bases in the rear of the French Army from which operations against Germany could be carried out without the necessity of flying over Holland, which would have been the case had raids been launched from RNAS stations in southeastern England.[10]

Though the idea of a retaliatory force was a direct product of the Zeppelin raids and the public reaction to them, the Admiralty and its air arm had been interested in long distance bombing since the beginning of the war. It is possible that naval airmen were temperamentally more receptive to bombing operations than their army counterparts in the Royal Flying Corps, and that to them independent and wide-ranging forays by bombers were closely akin to well-established forms of naval operations. At any rate, the RNAS had made a number of long-distance raids against Zeppelin bases in 1914, and had continued to emphasize bombing both during the Dardanelles campaign and from its base at Dunkirk. Winston Churchill, when First Lord of the Admiralty, had encouraged this policy, and as early as December 1914 he had authorized the design of a heavy bomber, the Handly-Page 0/100, which became operational in 1916. When he left office he maintained his interest in the possibilities of bombing, and on two occasions in early 1916 urged the bombing of Germany. On 17 May he told the House of Commons:

> The air is free and open. There are no entrenchments there. It is equal for the attack and for the defence. The resources of the whole world are at our disposal and command. Nothing stands in the way of our obtaining the aerial supremacy in the War but yourselves. There is no reason, and there can be no excuse, for failure to obtain that air supremacy, which is, perhaps, the most obvious and the most practical step towards a victorious issue from the increasing dangers of the War.[11]

These remarks, addressed specifically to the use to which the RNAS might be put by the Admiralty, underline as well Churchill's deep frustration with the Western Front deadlock. The parallel with his advocacy of the Dardanelles expedition is close; Churchill was perhaps the first major British politician to see in the bomber a means to escape the paralysis of trench warfare. And, like others, he was led to attribute to the air weapon an efficacy it was far from possessing.

During 1916 and early 1917, raids against German cities were carried out by 3 Wing, RNAS, from bases at Nancy, Luxeuil and Ochey. About ninety percent of the wing's air crew were Canadians. Among the targets they sought to attack in this first British bomber offensive were the Mauser factories at Oberndorf and steel works at Hagendingen, Dillingen and Saarbrucken. This was in accord with their orders, which laid down that objectives "should be of military value and promiscuous bombing of unfortified towns should on no account be permitted."[12] Though evidence from German sources is scanty, it is doubtful whether the RNAS was any more successful in hitting military targets than were the Zeppelins. A case in point was the raid of 27 December 1916 against the Dillingen blast furnaces by eleven bombers. Broken clouds and thick haze made identification of the target virtually impossible; nevertheless the wing's commander concluded that "as the target is a large one it is probable that many bombs reached the objective."[13]

The Admiralty undertook no analysis of the effectiveness of these raids. Instead, in the enthusiasm which followed them, naval authorities invited Colonel Bares of the French air service to London to discuss the expansion of the RNAS force and the launching of a genuine strategic bombing offensive. Such a development was adamantly opposed both by General Haig and by the Royal Flying Corps, partly because it would divert vitally needed engines and aircraft from the Somme air battle, which had reached a critical stage in the late autumn of 1916, but principally because strategic bombing was rejected in principle. Haig argued, in a strongly worded protest of 1 November, that victory in the war could only come about through the defeat of the enemy's forces in the field. To this end the prime aviation requirement was adequate numbers of artillery cooperation, photographic reconnaissance and contact patrol aircraft, together with the best possible fighters to protect them. "The next most urgent requirement," he wrote, "is reconaissance behind the enemy lines, and bombing of such railways, headquarters, bivouacs, etc., as may affect the issue of a battle by upsetting the enemy's organisation and command, and interfering with his tactical and strategical movements." Haig bluntly opposed long range or strategic bombing because it contributed nothing to the outcome of the ground battle, and because it was immoral.

> Long distance bombing as a means of defeating the enemy is entirely secondary to the above requirements. Its success is far more doubtful, and, even when successful, both theory and practice go to show that usually its results are comparatively unimportant.
>
> I have no reason to suppose that the bombing of open towns merely for the purpose of terrorising the civil population is a method of warfare which would be approved by His Majesty's Government, nor would I recommend its adoption.[14]

Even in its limited aim of drawing off German fighter strength from the Western Front, as Haig rightly observed, the RNAS bombing campaign had produced no discernible results. Haig's opposition, together with the air crisis over the Western Front and the hostility of the War Office towards an air offensive carried out by the Navy from bases in France, forced the Admiralty to abandon its plans for a new direction in the air war and caused the disbandment of 3 Wing in April 1917.

Within a few months the War Cabinet reached a position diametrically opposed that of Haig and made strategic bombing an official part of British war policy. This was brought about by the German heavy bomber offensive against England which began in mid-1917. The German High Command became convinced in late 1916 that the airships were in process of being bested by the British defence system. Without altogether abandoning attacks by Zeppelins, it determined to step up raids on England with the newly developed Gotha G.IV bomber. This twin-engined biplane had a ceiling of 21,000 feet, endurance of more than four hours, and could carry a bomb load of up to 500 kg., although in raids against England six 50 kg. bombs usually were carried. The operation, designated *Turkenkreuz*, was assigned to Kampfgeschwader 1, which flew from bases in Belgium. As laid down by OHL, the squadron's objectives were so to disrupt English industry and communications and strike at civilian morale as to "split up the numerically superior forces of the Allies in the air" by causing the withdrawal of RFC squadrons from France. In other words (having in mind Haig's critique of Admiralty intentions) the German High Command, in deciding to embark upon the *Turkenkreuz* campaign, calculated that strategic results on the Western Front could be obtained by an economical use of air power.[15]

On 25 May the Gothas first appeared in force over England. They bombed Shorncliffe Camp; 110 Canadian soldiers were among the casualties. On 13 June Kampfgeschwader 1 made its first raid on London, causing 600 casualties including 66 children killed and injured when a bomb struck an East End school. On 7 July the Gothas made their most spectacular raid on the capital. Appearing over London in mid-morning, 21 bombers dropped explosives over a wide area. Though casualties and damage were not heavy, the raid, viewed by millions of Londoners, vividly disclosed the impotence of a defense system designed to ward off airships, not airplanes.[16]

Through the periodic transfer of front line squadrons from France to England and the stiffening of home defense squadrons by allocations of first class fighter aircraft and experienced combat pilots, the British home defence organization forced the Germans to go over to night raiding by late 1917. These night raids, in which Germany employed Giant aeroplanes (*Riesenflugzeuge*) as well as Gothas, were never seriously impeded by the defenses until the final raid of the war in May 1918. At that time the Germans gave up the campaign of their own volition, deciding that the bombers could more directly serve the needs of the army in the field by operations against the Allied back areas on the Western Front.

The German raids achieved their purpose. A very elaborate air defense system had been built up in Britain, under the command of Brigadier General E.B. Ashmore, RAF, and the mere threat of the resumption of bombing was enough to keep it in being for the rest of the war. London Air Defence Area mustered eleven RAF fighter squadrons and three balloon squadrons; in the Northern Group there were five additional fighter squadrons. Thus, more than two hundred first class aircraft, their pilots, and administrative staff and ground crews (a far larger number of men) were tied down in Britain instead of being available for continental service. The number of men needed to operate the ground defenses against air attack, including headquarters staff, searchlight and sound-ranging crews, gunners, and communications staff, was 15,115.[17] When, in July 1917, Chancellor Bethmann-Hollweg protested the raids on England because they "irritated the chauvinistic and fanatical instincts of the English nation without cause," Field Marshal von Hindenburg was on strong ground when he replied:

> We must... prosecute the war with all our resources and the greatest intensity. Your Excellency deprecates the aerial attacks on London. I do not think the English nature is such that anything can be done with them by conciliation or revealing a desire to spare them. The military advantages are great. They keep a large amount of war material away from the French front and destroy important enemy establishments of various kinds. It is regrettable, but inevitable, that they cause the loss of innocent lives as well.[18]

Yet the Chancellor had a point. Anger in Britain mounted with each successive Gotha raid. At first the prevailing notes were confusion and frustration, and public wrath was directed not simply at the Germans but at the Royal Flying Corps, the home defence apparatus and the politicians who were responsible for it. Little or no censorship appears to have been exerted over the protests appearing in the press, nor did the authorities lack the benefit of a multitude of suggestions for dealing with the German menace. In *The Aeroplane*, C.G. Grey dismissed "the usual outcry about the Hunnishness of bombing women and children" as mere cant; instead, he suggested the evacuation of women and children from what had become a new war zone, or else the destruction of the raiders' bases in Belgium.[19]

After the initial German attack of 25 May, the War Cabinet considered responding in kind of the enemy's "frightfulness," but decided instead to patch up the home defense organization. Sir John French, Commander-in-Chief Home Forces, warned that such makeshift measures would have disastrous consequences; he was vindicated by the raid of 13 June. The War Cabinet then turned to General Trenchard for expert advice. In a memorandum promisingly entitled "Measures suggested for the preventing of air raids in the United Kingdom," Trenchard provided no comfort whatever to the politicians. He made short work of a suggestion that the RFC fly round-the-clock defensive patrols, pointing out that the number of aircraft required to make such a scheme effective went far beyond any foreseeable supply. No defense system, he declared, could ever prevent bombers from getting through. Nor were reprisal raids against German cities feasible, unless the bombers to make them were taken from the already short-handed RFC on the Western Front. German raids could only be countered by capturing their bases during the Flanders offensive or destroying them by sabotage.

The War Cabinet had no alternative but to accept Trenchard's advice. Nevertheless, in a series of remarkable decisions, it gave air production priority over all other weapons production, authorized the virtual doubling of the RFC from 108 to 200 squadrons, ordered the stepping-up of airplane engine production from 1,200 to 4,500 a month, and provided that forty of the RFC's new squadrons were to be set aside to carry out the reprisal bombing of German cities, commencing, in all probability, in the spring of 1918.[21] None of these measures had any immediate effect; meanwhile the German raids continued. When nearly one hundred RFC fighters failed to prevent the Gothas from bombing London on the morning of 7 July, the War Cabinet held an emergency meeting the same afternoon.

It is no exaggeration to say that the Cabinet was in a state not far removed from panic. "One would have thought the world was coming to an end," Sir William Robertson observed to Haig; "I could not get in a word edgeways." The outcome of this meeting was a Cabinet order to Haig to carry out a reprisal raid upon Mannheim, unless this would "completely disrupt his plans." When Haig pleaded lack of resources and made plain his skepticism of the efficacy of reprisals, the War Cabinet turned for advice to another quarter. On 11 July Lieutenant General Jan Christian Smuts, whose military qualifications had no connection with air matters, was named as a committee of one (though nominally the Prime Minister was to act with him) to look into two subjects, defense against the bomber and the whole question of British air organization and its bearing upon air operations.[22]

Smuts made a number of recommendations for strengthening home defenses, including the appointment of a senior air officer to take executive control of the system. This resulted in the appointment of Brigadier General Ashmore. It was Smuts' view, however, that defensive measures no matter how efficient would not be enough to stop German raiding, particularly night attacks. The whole subject was fraught with potential danger to the "nerve centre" of the Empire, especially if the Germans resorted to the use of gas bombs. His recommendations, he thought, touched "only the fringe of the danger"; the real answer to air attack was "offensive measures."[23] The future direction of British air policy was therefore at the heart of his important report on air organization, submitted to the War Cabinet on 17 August.

In compiling his report, Smuts had the advantage of advice from Lieutenant General Sir David Henderson, Director General of Military Aeronautics and the RFC's senior officer from its inception. Henderson seized the opportunity to express his exasperation with the division of responsibility between the RFC and the RNAS and to argue for the creation of a separate air service, independent of both Army and Navy, with its own Air Ministry and general staff. Though he conceded that to carry out such a reorganization in time of war might result in temporary loss of efficiency, he believed that the immediate benefits deriving from the end of duplication and wasteful rivalries and competition would be considerable, especially in matters of supply, training, research and development, and policy and operational planning. Smuts accepted these arguments and made the amalgamation of the RFC and RNAS, the creation of an Air Ministry, and the establishment of "an Air Staff on the lines of the Imperial General Staff, responsible for the working out of war plans, the direction of operations, the collection of intelligence, and the training of air personnel" his chief recommendations to the War Cabinet.[24]

Underlying Smuts' advocacy of an independent air force was his belief that the air had added a new dimension to war which the British, because of organizational confusion, had not yet adequately exploited. The major weakness was the failure to develop an air offensive against the German homeland. To Henderson, this was only one of the several considerations he had laid before Smuts:

> It is, of course, evident that until the immediate needs of the Navy and Army can be supplied, there will be no central Air Force available for independent operations. So far as the Army is concerned, however, there is reason to hope that its immediate need for fighting, for reconnaissance, and for artillery, work, will be met in the early months of next year, and that even then a considerable force of bombing machines will also be available. If the Air Ministry were in existence now, it would be its duty to look ahead and consider the best means of employing this Service, that is to say, considering for instance whether it could better be employed under the direct command of the Commander-in-Chief in France, or only under his nominal command, if serving in France, but strategically directed by the General Staff of the Air Ministry.[25]

For Henderson, then, the idea of a distinct strategic bombing force for "independent operations" was dependent upon prior satisfaction of the needs of the senior services, though he clearly envisaged the creation of such a force as feasible were a separate air arm to be formed.

Smuts was far less cautious. Not only did he give more weight to the optimistic projections of aircraft production made by the Ministry of Munitions, but he took a far more imaginative and sweeping view of the war than did Henderson. Indeed, he had persuaded himself that supremacy in the air might well be a key to victory. "It is important for the winning of the war," he contended, "that we should not only secure air predominance, but secure it on a very large scale." The War Cabinet had authorized a programme of aircraft production, in large part because of German air raids, "far in excess of Army and Navy requirements." But who was to determine the best use for this surplus? For himself, Smuts had no doubt about the form its employment should take, and who ought to direct it. "Independent strategical operations" should become the major responsibility of a new air organization.[26]

It was incorrect, he argued, to think of the air weapon in the same terms as artillery; that is, as an arm ancillary to military or naval operations. Beyond the services it could provide to armies and navies the air arm had a distinctive strategic potentiality.

> Air service . . . can be used as an independent means of war operations. Nobody that witnessed the attack on London on 11th July [Smuts meant on 7 July.] could have any doubt on that point. Unlike artillery an air fleet can conduct extensive operatins far from, and independently of, both Army and Navy. As far as can at present be foreseen there is absolutely no limit to the scale of its future independent war use. And the day may not be far off when aerial operations with their devastation of enemy lands and destruction of industrial and populous centres on a vast scale may become the principal operations of war, to which the older forms of military and naval operations may become secondary and subordinate.[27]

The ease with which Smuts moved from an unexceptionable appreciation of the strategic possibilities inherent in the air weapon to an apocalyptic vision of its future is typical of air power enthusiasts from that day to this.

Moreover, like air power enthusiasts to come, Smuts had a marked tendency to allow his vision of the future to affect his judgment of the current actuality, a species of mental confusion that was shared by other members of the War Cabinet. In relating his findings to the likely course of operations in 1918, Smuts was manifestly swayed by the very scale of air strength which, he believed, would be available to unleash upon the Germans.

> The magnitude and significance of the transformation now in progress are not easily realized. It requires some imagination to realize that next summer, while our Western Front may still be moving forward at a snail's pace in Belgium and France, the air battle-front will be far behind on the Rhine, and that its contin-

uous and intense pressure against the chief industrial centres of the enemy as well as on his lines of communication may form an important factor in bringing about peace. The enemy is no doubt making vast plans to deal with us in London if we do not succeed in beating him in the air and carrying the war into the heart of his country.[28]

The contrast Smuts drew between the toiling armies on the Western Front and an air fleet vaulting over the stagnant war of the trenches, winning decisions almost bloodlessly and at a distant remove from the battlefield, appears to have made a powerful impression upon the War Cabinet. Smuts' plan offered an escape from "the progressive exhaustion of the man-power of the combatant nations," an opportunity to seize victory by exploiting "to the utmost" the industrial superiority of the Allies. "Mechanical power" was the way out of the impasse of the trenches; just as the submarine had wrought a startling transformation in naval warfare, so the aircraft was "destined to work an even more far-reaching change in land warfare."

On 24 August the War Cabinet accepted Smuts' recommendations in principle and established a committee to make the necessary plans and preparations. Its decision, however, was vigorously attacked by the two officers upon whom the Cabinet had previously relied for professional advice on military air operations, Haig and Major General Hugh Trenchard, the commander of the RFC in France. "The contention on which the whole argument for a separate Air Service is based is that the War can be won in the air against on the ground," Trenchard wrote. "Nothing but bare assertion is urged in support of this contention. It is, in fact, merely an opinion." Not only had Smuts overlooked the operational difficulties of bombing Germany, but he imagined that raids upon German cities and industries would bring strategic results despite the fact that enemy raids upon England "have had no effect whatsoever on the course of the war." A separate air ministry with a civilian head no longer checked by traditional sources of professional military and naval advice, Trenchard contended, would be unduly swayed by purely political considerations. It would be subject to "popular and factional clamour" and "drawn towards the spectacular, such as bombing reprisals and home defence, at the expense of providing the essential means of cooperation with our Naval and Military Forces."[29]

Haig was not so openly contemptuous of the motives of the War Cabinet and he was prepared to admit that long range bombing of German communications and naval and military resources "may certainly give valuable results." In addition, the bombing of cities as reprisal might have some justification, were it likely to bring German bombing to a halt. "Once such a contest is commenced, however, we must be prepared morally and materially to outdo the enemy if we are to hope to attain our ends." In Haig's judgment Britain did not have the resources to mount and sustain such an offensive, and he was clearly apprehensive that the results of the new policy would work to deprive the Army of the air strength it needed on the Western Front. Both officers argued that Smuts had grossly overestimated the air arm's capacity to produce strategic results; as Haig observed, the Smuts report gave a weight to bombing that officers of "wide practical knowledge" could not accept. The real danger was an Air Ministry "assuming control with a belief in theories which are not in accordance with practical experience."[30]

It was in vain, however, that Haig urged reconsideration. Though support for his position existed, his stature as one of the Cabinet's principal professional advisers had been eroded by the evident failure of his Flanders offensive, still mired short of Passchendaele. To most ministers, an expedient that might conceivably end the carnage in France, or at least quell the public outcry for effective measures against German bombing, was too attractive to resist. Waverers were brought into line by a series of night raids on England from 2 to 4 September, culminating in an attack upon London on the 4th. The following day the War Cabinet resolved that "we must carry the aerial war into Germany, not merely on the ground of reprisal." More German raids in the last week of September brought a renewal of public pressure upon the government to take action. On 1 October, over the last ditch objections of Haig and Trenchard, orders were given to detach two bombing squadrons from the British Expeditionary Force as the nucleus of a force to undertake, as soon as possible, "a continuous offensive, by air, against such suitable objectives in Germany as can be reached by our aeroplanes."[31]

The organizational corollary to the new policy soon followed. On 6 November the Cabinet approved the Air Force Bill, and after passage through Parliament it was given royal assent on the 29th. Though the Royal Air Force would not come into being until 1 April 1918, the Air Ministry and the Air Staff were set up in January to pave the way for the new service and to plan its independent operations against Germany. Both the Secretary of State for Air, Lord Rothermere, and the Deputy Chief of Air Staff, Rear Admiral Mark Kerr, were committed believers in a bombing offensive, which made their relations with Trenchard, appointed Chief of Air Staff, uncomfortable from the start. Many differences, political and personal, divided these men, but flat disagreement over bombing policy was important among them.[32]

Trenchard's position was straightforward. He continued to believe, as his subsequent conduct of bombing operations was to demonstrate, that the first priority of air power was to serve the Army. Beyond this, he found the new air staff unduly optimistic about bombing, and his minister impervious to advice. He therefore took the step of communicating directly with the Prime Minister. He informed Lloyd George on 13 January that "There appears to be in some quarters very serious misapprehension as to the extent to which Germany can be bombed during the spring and summer of this year." Challenging staff forecasts that forty squadrons would be available for this purpose by the beginning of July, Trenchard offered his own estimate that no more than nine would be ready by that date, and perhaps thirty by the end of October. As he suggested to the Prime Minister, "It is far better to know what can really be done . . . than to indulge in more generous estimates that cannot be realized." With the small force he anticipated, he considered that "substantial mischief" could be done to industrial centres on the Rhine and to such nearer industrial targets as Saarbrucken.[33]

"Substantial mischief" was a far cry from the devastation Rothermere and those around him hoped to achieve, and the cold water Trenchard threw upon their grandiose dream of a mighty air fleet attacking Germany was most unwelcome. The kind of scenario Rothermere wanted to hear was typified by that provided by Sir Henry Norman, M.P., a former journalist whom the Minister had appointed to the Air Council. Norman called for the utter obliteration of Essen, Cologne, Mannheim, Stuttgart, Dusseldorf and Frankfurt, which he believed could be accomplished by a force of 250 bombers attacking each city in turn in relays of 25 bombers an hour. Raids of ten hours' duration would overwhelm urban defences and services, and as a result, these key cities would be "practically wiped out, so far as their collective existence and productivity were concerned." His opinion was that "if such attacks were pursued for a month, our victory in the war would be in sight." Doubtless Norman was one of those Trenchard had in mind when he complained to Haig that Rothermere had "introduced a lot of people to the Air Ministry without consulting anybody," and that the Minister "preferred any advice to that of his professional advisors."[34]

Before the Royal Air Force was born, Rothermere, Trenchard and Kerr had all resigned. Lord Weir of Eastwood became Secretary of State; his Chief of Air Staff, an old service rival of Trenchard, was Major General Frederick Sykes. Both Weir and Sykes unquestioningly accepted the strategic bombing role for the RAF, as did the planning staff they assembled. Sykes directed the attention of his planners to what he termed the "root industries" of Germany. On the basis of the economic intelligence gathered by the Admiralty's staff, he estimated that as much as eighty percent of the

German chemical industry and ninety-five percent of the magneto industry was within reach of British bombers flown from French bases, and he suggested that it was conceivable that twelve raids might account for this portion of the chemical industry, and another three might obliterate the magneto industry. Such an outcome was predicated upon the calculation that a chemical works could be destroyed by a thousand-sortie raid, and that five hundred sorties would be required to level a typical magneto factory.[35] Though it must be stressed that these were planning hypotheses, they bore no relationship whatever to the realities of the air war. Even if Sykes had been right and German factories could have been destroyed with the force he outlined, a minimum of 13,500 sorties would have been required. As it happens, the bomber force under his overall direction was unable, during the whole course of its operations, to fly as many as 3,000 sorties, and never carried out more than 45 in any 24-hour period between June and November 1918.

Victory through the destruction of German industry was nevertheless the principle behind Sykes' creation of the Independent Air Force (IAF). This strategic bombing force had as its nucleus the Nancy squadrons, now knows as VIII Brigade RAF, which had been raiding German industrial centres since late October. The IAF was to operate directly under the Air Staff in London, independently of the RAF command serving the British armies on the Western Front. Trenchard, whose lack of employment was a source of embarrassment to the government and to the service he had done so much to shape, was prevailed upon to take command. Sykes gave him the general objective of "the demobilization of the German Armies-in-the-Field, by attacking the root industries which supply them with munitions." He was forbidden to use his force against enemy airfields or communications, targets which properly belonged to the RAF's main force; were the IAF to attack such objectives, Sykes warned, "no real progress will be made in the destruction of key industries." Trenchard's first priority was to "obliterate" the German chemical industry; not until it had been "completely crippled" was he to turn to steel and other industrial targets.[36]

This survey of the development of a strategic bombing policy is not concerned with the actual operations of the IAF from Trenchard's assumption of command in June until the end of the war. Its operation, however, were a grave disappointment to the Air Staff and to those politicians who had pinned high hopes upon bombing. The IAF was hardly the air fleet envisioned by Smuts and the War Cabinet; nevertheless, many of the reasons for its failure demonstrate how deluded the government had been by the promise of easy victory through air power. The advice tendered by Haig and Trenchard, ignored though it was, had been sound in every particular, stemming as it did from actual knowlege of the air weapon and its limitations.

When Trenchard took over the IAF he found only five squadrons. He was promised seven more by mid-July and told to expect another 27 at intervals thoughout the summer and autumn.[37] Had he received all he was promised, his force would still have been far too small to carry out its orders. His peak strength, reached in the fall, was ten squadrons. The operations of this little force, conducted with great courage and resolution by its air crews, revealed the shortcomings both of men and machines. Relatively untrained crews were called upon to make long flights in slow aircraft, sitting in open cockpits fully exposed to cold and weather, and subjected to anti-aircraft fire and fighter attack, and despite the fatigue and nervous strain induced by these factors, to locate individual factories and bomb them accurately from heights of several thousand feet. It is hardly surprising that their real successes were few and their losses heavy.

To find the target city, let alone the factory it contained, was difficult enough; for night bomber crews it was often impossible. Navigational aids were rudimentary; most crews used dead reckoning supplemented by their knowledge of country and what they could add to the limited amount of meteorological information supplied them at the outset of a raid. Most day bombing squadrons were equipped with the De Havilland 9 or 9a; both aircraft had so low an operational ceiling and so little speed that they were highly vulnerable to fighter attack. As well, they were subject to repeated mechanical failures, and many of the crews lost were compelled to make forced landings within German lines. Only the Handley-Page 0/400, a fine night bomber, was up to its job. Bomb sights, though considerably improved from the earliest models, were far from satisfactory, and many bombardiers appear to have trusted to the eye alone. Although crews frequently reported direct hits upon their targets, many reports merely note that bomb loads were dropped "in the centre of town," a usual occurrence when the bombers were beset with fighters while over the target. Finally, the bombs used were much too light to do significant damage except with direct hits. Only towards the end of the war was a 1600-lb. bomb employed, and Handley-Page crews did obtain some noteworthy results with it.

In spite of his orders, Trenchard devoted a considerable part of the IAF's operations to attacks upon enemy airfields and communications, which he justified to Weir as necessary to prevent German bombing of his own bases, or as essential support for the ground operations of the Allied armies. Though Trenchard's divergence from the main plan was much criticized within the Air Staff, it does not appear that Sykes attempted to assert his authority and bring the IAF commander into line.[37]

Intelligence appreciation of bombing results was not highly developed during the First World War. As early as June, howver, Air Staff analysis of air photographs of damage caused by IAF raids disclosed that results were much short of expectations. It was found, for example, that only 23.5% of the bombs dropped fell within the target area—an estimate that was almost certainly a gross exaggeration. Some staff planners, impervious to evidence of this kind, rejected the "error curve" in bombing. One such response illustrates the fantasy world they inhabited:

> Putting aside all error-curves and coming down to simple arithmetic, it is a considerable underestimate to suggst that a 230 lb bomb will obliterate everything within 10 yards of it. Assuming this, however, one could divide an area into squares of 20 yards by 20 yards, with a bomb in the middle of each. Krupp's works and workmen's dwellings would contain 7,744 such squares. It is proposed to drop in one effort 20,000 such bombs. . .If Krupp's works are not obliterated by such an attack, then there is something very peculiar about them.[38]

Gradually, however, most of the planning staff turned away from the strategy of precision bombing of industrial targets, which they were compelled to recognize as beyond reach technologically, to the idea of striking a the will of the German population through terror attacks. In a paper prepared for the Cabinet, Sykes pointed out the advantages of what one day would be called area bombing. "The wholesale bombing of densely populated industrial centres," he observed, "would go far to destroy the morale of workers." When, in the course of the summer, it was realized that an efficient small incendiary bomb would soon be available, the staff compiled lists of German towns containing large stocks of working class housing. It was suggested that incendiaries might best be employed during daylight hours, "when the people are out at work, and perhaps only children left to look after the house." "I would very much like if you could start up a really big fire in one of the German towns," Lord Weir wrote to Trenchard, and assured him that he did not need "to be anxious about our degree of accuracy when bombing railway stations in the middle of towns. The German is susceptible to bloodiness, and I would not mind a few accidents due to inaccuracy."[39]

Thus the limitations of the air weapon had driven the chief advocates of its strategic employment to discard much of their theorizing, and to resort instead to the crude notion of victory through the mass slaughter and terrorization of civilians. Months before, while the Cabinet was still considering whether to approve the Air Force Bill, Winston Churchill, the

Minister of Munitions, had prophetically attacked the idea of terror bombing in a statement that surely has relevance for the Second as well as the First World War:

> It is not reasonable to speak of an air offensive as if it were going to finish the war by itself. It is improbable that any terrorization of the civil population which could be achieved by air attack would compel the Government of a great nation to surrender. Familiarity with bombardment, a good system of dug-outs or shelters, a strong control by police and military authorities, should be sufficient to preserve the national fighting power unimpaired. In our own case we have seen the combative spirit of the people roused, and not quelled, by the German air raids. Nothing that we have learned of the capacity of the German population to endure suffering justifies us in assuming that they could be cowed into submission by such methods, or, indeed, that they would not be rendered more desperately resolved by them.[40]

These views, as we have seen, were also those of Haig and Trenchard. It does not appear from the operational orders of the IAF that Trenchard paid the slightest attention to the entreaties of the Air Minister or to the shift in emphasis on the part of the Air Staff.

Neither the minister nor the Air Staff gave up the idea of striking some spectacular blow that would demonstrate the power of the air weapon and play a part in bringing Germany to its knees. From July 1918 on, their attention increasingly was drawn to the Handley-Page V 1500, an immense four-engine biplane which, under ideal conditions, could carry a bomb load of more than three tons. Its Rolls-Royce Eagle VIII engines gave it a range of more than a thousand miles, which made the bombing of Berlin from English bases a real possibility. While waiting for factory- and flight-testing to be completed, therefore, the RAF formed a special unit of the IAF, called 27 Group, to fly the V 1500 and to plan the operation against the German capital. It was given a base in Norfolk and put under the command of an experienced Canadian airman, Colonel W. R. Mulock of Winnipeg.[41]

As it happened, the V 1500 was not ready for operational use until early November, and although Mulock was kept in readiness until the last moment, the huge bomber never flew over Germany. However, the staff developed a number of scenarios for its employment, chiefly as a terror weapon. It was calculated, for example, that the Handley-Page could carry 16,000 small incendiary bombs, and that such a load, if correctly delivered, would lay down a belt of fire 60 yards wide and 2,500 yards long. In the words of the planner, "If the target is large the operation may be described as simply a plastering of the locality with a predetermined density of fire nuclei." It was thought, however, that the greatest effect upon built-up areas could be obtained by a combination of high explosive and incendiaries: "To obtain the maximum strategic, moral and material effect from an attack upon a suitable target such as a town...the effect of depositing high explosives closely followed by incendiary bombs could hardly be improved upon. The results might safely be described as terrific..."[42] To a remarkable degree, the thoughts of these pioneer strategic bombing planners foreshadowed those of RAF planners a generation later, through a direct relationship seems unlikely.

In another sense, however, the planning staffs of the two wars are inextricably linked. The chief raison d'etre of the RAF, from its inception, was strategic bombing. There were several reasons why a new service was created in 1918. From the early days of the war, there had been costly and acrimonious competition between the flying services. In addition, from the onset of the Zeppelin raids, public dissatisfaction with the lack of coordination in home defense arrangements had been building up. But neither of these factors was sufficient to persuade the government to carry out the wholesale transformation of its air organization in time of war. Moreover, as Ashmore's innovations demonstrated, an effective revamping of home defense could be carried out without the necessity of amalgamation, even though the structure he created was a distinctive kind of air and ground force.

It was the coming together in late 1917 of a number of developments, some short-term and accidental, others more fundamental, that made the case for a new service compelling to a majority of the Cabinet. For one thing, aviation was now to be taken seriously. From being an almost comic addition to the traditional fighting arms, the air weapon had emerged as an important and often crucial element, especially in the land battle. There was a growing sense, not only among airmen but among politicians and informed members of the public, that a divided and subordinate flying service made impossible the full exploitation of the new weapon. The Gotha raids not only intensified the public demand for a more aggressive air policy, but they provided, at least for those not disposed to be overly critical, a vivid demonstration of what air power could do. Heretofore, advocates of the strategic use of air power had been overborne by the weight of adverse professional opinion, especially that of Haig and Trenchard. At this juncture, however, Haig was seen as the chief architect of failure on the Western Front; not only did his advice lack the authority it had once commanded, but his position was in some jeopardy. It was Smuts, the brilliant fixer, who condensed these and other tendencies into the eloquent reports upon which the Cabinet based its decision. At this stage in the war, the idea of an alternative to the carnage of the trenches must have ben infinitely attractive to some ministers, and worth the gamble to others. Moreover, though politicians, for the most part, had resisted the call for reprisal, they were not immune to the deeply felt desire for revenge among broad sections of the British public, not only because of German raiding but because of the conviction of German guilt for the war. Bombing would visit upon the German people the horror of the war for which they were held to be responsible, and from which, to this point, they had been shielded.

That the bombing campaign bore no relation to the rhetoric of politicians nor to the visions of planners is less significant than the association it established between the RAF and strategic bombing. When, immediately after the war, a British commission investigated the results of IAF raids, it concluded that though physical damage had been minor, nevertheless the effects on German morale were such that "had the war continued a few months longer, a more or less total breakdown of labour at several of the Works might have been confidently expected." Sykes' contribution to the same myth was his assertion that had the bomber force been augmented to 500 aircraft, "there can be no reasonable doubt that the Germans must have collapsed during the summer of 1918. . ."[43] There was no real basis for such judgements. But the groundwork for the defense of the RAF as a distinct service was being laid, and Trenchard himself used such arguments when he returned as Chief of Air Staff in 1920. The consequence was to bias the RAF towards a bombing strategy, at the cost both of the fighter arm and of the painfully-won expertise gained when the flying services had operated as ancillary arms of the Army and Navy. Moreover, the bombing mystique derived from the First World War was accepted uncritically, without clinical analysis of the shortcomings of the IAF. As a reviewer in the *Times Literary Supplement* observed concerning the state of the RAF in 1937, "the whole raison d'etre of the RAF as an independent service was based on considerably less research than would normally be regarded as adequate preparation for a wager on a horse race."[44] The same could be said for the RAF's first strategic bombing campaign in the First World War.

Notes

1. T.H.E. Travers, "Future Warfare: H.G. Wells and British Military Theory, 1895-1916," in Bond and Roy, eds., *War and Society*, London, 1975, 74ff.
2. D.H. Robinson, *The Zeppelin in Combat: A History of the German Naval Airship Division*, London, 1962, 50.
3. Quoted in *ibid*, 67.

4. H.A. Jones, *War in the Air*, III (Oxford, 1931), 165-167.
5. *Ibid*, 188; London, Public Record Office, Air 1/574/16/15/160; Air 1/573/16/15/156.
6. *War in the Air*, III, 135-136, 141-142; Robinson, 124, 127-128; Air 1/573/16/15/156; London *Daily Telegraph*, 18 March 1970.
7. House of Commons, *Debates*, 16 February 1916; *The Aeroplane*, X, February 1916, 324-326.
8. Air 1/582/16/15/181/Pt. 3; Robinson, 173-175.
9. Air 1/300/15/226/151/Pt. 1.
10. S.W. Roskill, *Documents Relating to the Naval Air Service*, London, 1969, I, 334; "History of 3 Wing, RNAS," n.d., Air 1/648/17/122/397; Minutes of 28th meeting of Air Board, 1 November 1916, Air 6/3.
11. House of Commons, *Debates*, 17 May 1916.
12. Admiralty to O.C. 3 Wing, 27 July 1916, Air 1/300/15/226/151/Pt. 1.
13. Wing Commander Elder to Admiralty, 28 December 1916, Air 1/648/17/122/397.
14. Haig to War Office, 1 November 1916, Air 2/123/B10620.
15. R.H. Fredette, *The Sky on Fire: The First Battle of Britain, 1917-1918, and the Birth of the Royal Air Force*, New York, 1966, 34-37; Peter Gray and Owen Thetford, *German Aircraft of the First World War*, London, 1962, 128-132; Ernst von Hoeppner, *Deutschlands Krieg in der Luft*, Leipzig, 1921, 112.
16. Air 1/588/16/15/197; von Hoeppner, 111-112; Minutes of 100th meeting of Air Board, 14 June 1917, Air 6/8; Air 1/589/16/15/199; Air 1/590/16/15/202; Air 1/646/17/122/360.
17. E.B. Ashmore, *Air Defence*, London, 1929, 93-94, 108-109; *War in the Air*, V. Oxford, 1935, App. IX, "Anti-aircraft defences in Great Britain at the Armistice," 505-507.
18. Erich von Ludendorff, *The General Staff and Its Problems*, New York, 1919, II, 452, 497.
19. *The Aeroplane*, XII, May 1917, 1364.
20. Minutes of 154th meeting of War Cabinet, 5 June 1917, PRO Cab 23/3; Minutes of War Office conference, 31 May 1917, Air 1/614/16/15/318; War in the Air, V, App. IV, "Methods suggested for preventing air raids in the United Kingdom," 479-482.
21. Minutes of 154th and 169th meetings of War Cabinet, 5, 26 June 1917; Air 2/87/304.
22. Sir William Robertson, *Soldiers and Statesmen*, London, 1926, II, 17; Minutes of 178th meeting of War Cabinet, 7 July 1917, Cab 23/3; *War in the Air*, V, 38-42; W.K. Hancock, *Smuts: the Sanguine Years, 1870-1919*, Cambridge, 1962, 438-442; D. Lloyd George, *War Memories*, IV, London, 1934, ch. 68.
23. "First report of the committee on air reorganization and home defence against air raids," Air 1/609/16/17/275.
24. *War in the air, Appendices*, Oxford, 1937, App. I, "Memorandum on the organization of the air services," 19 July 1917, 1-8; App. II, "Air organization," 17 August 1917, 8.
25. *Ibid*, App. I, 5.
26. *Ibid*, App. II, 10-11, 14.
27. *Ibid*, 10.
28. *Ibid*, 11.
29. *War in the Air*, VI, Oxford, 1937, 13; Trenchard to Kiggell, 30 August 1917, Air 1/521/16/12/3.
30. Haig to Robertson, 15 September 1917, Air 1/521/16/12/3.
31. Minutes of 228th meetings of War Cabinet, 5 September 1917, Cab 23/4; Robertson to Haig, 1 and 2 October 1917, Air 1/970/204/5/1108; Minutes of 243rd meeting of War Cabinet, 2 October 1917, Cab 23/4.
32. Mark Kerr, *Air, Sea and Land*, London, 1927, 284-294; Andrew Boyle, *Trenchard, Man of Vision*, New York, 1962, 245-270.

33. Trenchard to Lloyd George, 13 January 1918, Air 1/522/16/12/5.
34. Norman to Rothermere, 25 March 1918, Air 1/2422/305/18/17; Boyle, 261.
35. "The formation of a strategic committee dealing with air matters," 19 April 1918, Air 1/450/15/312/4.
36. "Operation order for the guidance of the Independent Force," n.d., Air 1/462/312/16; Frederick Sykes, *From Many Angles*, London, 1942, 544-554.
37. Trenchard to Weir, 2 July 1918, Air 1/2000/204/273/275; P.R.C. Groves to Sykes, 11 September 1918, Air 1/460/15/312/97; also "Squadrons allotted to the Independent Force, Royal Air Force," 13 May 1918, Air 1/30/15/1/155/1-3.
38. Major Tiverton to Director of Flying Operations, 22 June 1918, Air 1/461/15/312/107.
39. Sykes, 550-551; Tiverton to Brig. Gen. P.R.C. Groves, 11 June 1918, DFO File, Directorate of History, National Defence Headquarters, Ottawa; Boyle, 312.
40. *War in the Air, Appendices*, "Munitions possibilities of 1918," 21 October 1917, 19.
41. Mulock Papers, DHist, NDHQ, FO3 to DFO, 4 July 1918, Air 1/461/15/312/107.
42. "Incendiary operations as a means of aerial warfare," 30 September 1918, Air 1/461/15/312/111.
43. "Survey of damage," Air 1/1999/204/273/269; 16; Sykes, 231-232.
44. *Times Literary Supplement*, 30 June 1972. The reviewer probably was Michael Howard.

THE BOMBING WEAPON

The Development of Air Raid Precautions in World War I

Marian C. McKenna
University of Calgary

The appearance in the skies over London of German airships and airplanes in World War I represented the implementation of an earlier "theory of frightfulness" advanced in the first decade of the twentieth century by the Italian General Giulio Douhet, one of the first military officials to have the foresight to perceive the war potential inherent in those untried contraptions called airplanes. Drawing from classic military strategy as it was then taught, Douhet deduced that the air arm should have for its essential function the weakening of key defensive points in the enemy lines by bombing. But behind the lines there would be other targets—rear bases, supply routes, lines of communication, factories, and civilian populations in cities large and small. The air corps was not to be used as a mere tactical tool of the army or navy, but was to operate as an independent unit, a separate striking force. The inevitability of clashes between opposing aircraft, and the bombing of cities from the air formed an important part of Douhet's conclusions.[1] In airing these beliefs, the General lost almost all his friends in the Italian military. Detractors mocked his "theory of frightfulness." For suggesting that urban centers be subjected to aerial bombardment he was labeled a barbarian.

Some of Douhet's ideas about aerial warfare were not entirely new, but had been the dream of earlier visionaries. Balloons had been used for observation by French armies in the late eighteenth century. In the United States Colonel Thaddeus Lowe's Union Army Balloon Corps performed effectively during the Civil War in reconnaissance, particularly in the Eastern theater, above the battlefields of Virginia.[2] An aeronaut veteran of this corps, John H. Steiner, a German-American who left the service in a dispute over back pay, went to St. Paul, Minnesota, where he earned a living taking passengers aloft for $5 apiece, and giving high altitude demonstrations. One customer was Count Ferdinand von Zeppelin, a young German lieutenant who visited the United States in the spring of 1863 to observe Civil War battles. After Zeppelin had seen enough of Union Army operations, he joined in a hastily planned expedition to explore the Mississippi River to its source. The expedition collapsed, but Zeppelin, staying with his party at a St. Paul Hotel, noticed an advertisement for one of Steiner's balloon ascensions. After a brief conversation (in German) between the two compatriots, an ascent was arranged which was to leave a deep impression on the young Count. Zeppelin wrote long afterward: "I wanted a real sensation and arranged for the use of the balloon, he to cut the rope and let me make a long flight after I had gotten up to the limit of the rope." He reports: "I was able to get up several hundred feet, but the gas was of such poor quality that I could not get the bag filled sufficiently to essay a long flight."[3]

Such was the first balloon ascent of the man who was to become synonymous with that of the dirigible he invented. The idea of such an airship took a strong hold on the imagination of German army and naval officers long before the outbreak of the First World War, and Count Zeppelin devoted the best years of his life to its development. From the beginning he met with great difficulties. The first ships were crippled by mechanical failures. After these were overcome, a series of freak accidents almost put an end to Zeppelin experiments. Through popular subscription and with government support, Zeppelin was able to continue, and by the time war broke out in 1914 Germany had about thirty-five dirigible balloons of his and other types, some more than four hundred feet in length, ready to be used in warfare.

The military potential of these airships was not understood by the Germans. Although Douhet had offered a wealth of shrewd prediction in his essay, "Rules for the Use of Aircraft in War," published in 1909, scarcely anyone in his or any other government accorded the work serious consideration. Even when bombs began to fall on England airships were not regarded as of great military significance. Nevertheless, before the end of the war, progressive military thinkers would be ready to adopt Douhet's theories as precepts of grand strategy. To a rising generation, the possibilities of air power in warfare opened up an entirely new philosophy of international conflict which would find its first major expression in the events following 1939. Even before the hostilities in World War I had ceased, some of the basic strategies and tactics of aerial warfare had been established. And when, during World War II, military aviation finally reached maturity, the grim predictions of General Douhet became a terrifying reality. Nevertheless, developments during World War I have been accorded little attention by military historians. Much of the German archival material was destroyed by Allied bombing in World War II, and not until recently has the wealth of official documents concerning aerial combat in World War I as seen from the British side been opened.[4]

Its inventors seem to have realized the airship's military possibilities. For some time before Douhet's essay appeared, Zeppelin had been trying to develop a balloon that could be propelled against the wind and so guided that explosives could be dropped upon hostile armies. Military plans were of course kept secret and such uses were repeatedly denied by the Germans, but technological advances along these lines were in a remarkable state of readiness by 1912. When war came the great aerial warships were given a thorough trial by the Germans.

The doctrines of unconditional warfare were demonstrated most dramatically in Zeppelin raids on France and England. The spectre of airborne attacks on cities had been delineated with gruesome detail well in advance of the outbreak of fighting, not only in the treatises of Douhet, but in the pages of the British press, yet in 1909 the British military chiefs scoffed at the notion that England could be vulnerable to assault from the air. So remote did the prospect seem that the War Office and the Admiralty did absolutely nothing to offset the French lead in heavier-than-air craft or the predominance of lighter-than-air ships of Germany, where Zeppelin had already emerged as a prophet vindicated, and where Pan-Germanists like General Friedrich Bernhardi were writing books foretelling of raids against Paris and London by mammoth dirigibles gliding mercilessly through the night skies.[5]

In the first months of the war German aerial operations got off to an inauspicious start. A key position in the way of the German army moving through Belgium was Liege. On August 6, 1914, the Zeppelin LZ-9 was sent from Cologne to bomb the city. The thirteen bombs dropped by the LZ-9, resulting in the deaths of nine civilians, inaugurated a new twentieth century practise.[6] Unfortunately for the Germans, however, within forty-eight hours three other Zeppelins were destroyed on their maiden missions.[7]

As a result of the losses, the Germans kept the recently commissioned LZ-9 in hangar for the next three months. Then followed an event which brought into clear focus the British fear of German airships. Antwerp, already battered by an artillery siege, was bombed. In the raid twelve civilians were killed, many more were injured, and part of a hospital was demolished. More raids followed by Zeppelins equipped with newly improved shrapnel and incendiary bombs. After each raid the British public read the details in their newspapers with growing alarm and foreboding.

Greater refinements characterized the newer airships built for the German Navy under the supervision of Captain Peter Strasser, head of the Navy's airship division after mid-1913. They were smaller in size than the Zeppelins to insure easier ground handling, and were equipped with dependable, heavy-duty Maybach engines; their radios were powerful enough to transmit and receive messages over a 300 mile range, so that meteorological data could be exchanged between bases and airships. Strasser worked for months to create an effective fleet of airships to bomb the British Isles. By December, 1914 his fleet was ready to strike, although the German Army, still shaken by its August losses, refused to cooperate with the Navy's proposal to combine and consolidate the two airship forces under a single command against the British.

The air war over England opened, not with a Zeppelin raid, but with an unexpected attack in December, 1914, when a single bomb dropped from an FF-29 seaplane fell harmlessly into a Dover garden. Another German raider visited southern England on Christmas Day. After making a reconnaissance of defense establishments and military targets beyond the Thames estuary, the plane recrossed the Channel, surviving attacks by Royal Naval Air Service pilots. The British, who had been expecting the worst, were more relieved than angered. Neither the British public nor the Germans were yet able to grasp that whatever the Zeppelins could do in scouting or in bombing could be better accomplished by airplanes. It was not airplanes but Zeppelins, with their heavier bomb loads, that the British public most feared. Military experts in England and France, however, had decided before the war that the airplane was the more deadly weapon, and in the end they were correct. The few thousand civilian casualties from combined Zeppelin raids in the early years of the war cannot be shrugged off as of no consequence, but neither were they cause for "runaway hysteria."[8]

The Germans eventually learned through hard experience that their fleet of airships and Zeppelins was not as effective as squadrons of Gotha airplanes, but they did not want to admit that they had made a costly mistake in adding the dirigibles to their armament. They could not give them up easily. Some were determined to use them in attacks far removed from the front lines, dropping bombs not on military installations where they might meet resistance, but on civilians in city streets.

In the late summer of 1914, German military leaders considered plans for breaking British resistance with a combined assault by sea and air. At that stage, however, the concept of sending bombing raids against England was still a subject of grave controversy between German militarists and the conscience of the moderates, as represented by Kaiser Wilhelm II and the German Chancellor, Theobald von Bethmann-Hollweg. A stumbling block in the realization of the "policy of frightfulness" was the Kaiser, who at first, more insistently than the Chancellor, deprecated any military course that would mean attacks on women and children, or on priceless landmarks and treasures of art and architecture. He went so far as to withhold approval from German Imperial Staff plans to conduct an offensive against military targets in England. One by one, Admirals Bachman and Pohl, the immediate subordinates of Admiral von Tirpitz, sent memoranda to the Kaiser showing that they shared none of his humanitarian considerations.[9]

Since August 23, the head of the German Naval Cabinet, Admiral von Muller, and the Supreme Naval Commander, Admiral von Tirpitz, had been discussing preparations for a combined air and naval attack on England. These ideas, known also to some diplomats, were outlined in a letter by the German Minister in Stockholm, von Reichenau:

> The war must be carried into their own land in every possible way, the population kept in continuous quaking terror. For that reason I hope with all my heart that we shall occupy and keep occupied the whole northern coast of France and Belgium from Cherbourg to Ostend, that there we shall build airship hangars at a safe distance from the sea, and send airships and aircraft cruising regularly over England and dropping bombs, and that we shall make continuous attacks against the English coast.[10]

Ultimately, the German Staff brought enough pressure on the Kaiser to permit limited air attacks on specified coastal targets, in return for solemn promises that only military objectives would be struck. Naval and military facilities along the banks of the Thames thus came within the sphere of legitimate assault by January, 1915, although the Kaiser was still unable to make up his mind whether or not to include London among the places to be attacked. This hesitation was shortlived. Following scattered night raids on England by airships of the German Navy, which roamed aimlessly over the eastern countryside, gaining little or no military advantage, new orders were issued permitting raids on the London docks, shipyards, armories, and other military targets. Attacks were to be scrupulously limited to these targets. Since the bombs were thrown by hand from airships or planes, and in view of the total lack of bombsights, it was a foregone conclusion that these raids would inevitably endanger the lives of civilians in large areas of central London.

The danger became a reality when the German Army, anxious to demonstrate the superiority of its airships over those of the Navy, briefed its commanders to attack any likely target in East London, without specifying waterside locations. Only bad weather delayed the onset of these attacks. After making token objections, the Kaiser finally relented and signed a directive permitting air raids after May, 1915 on sites east of the Tower of London. The next day, May 31, the first bombs were dropped on the East End.[11]

The Navy, embittered at having lost the initiative in this internal struggle, tried to persuade the Kaiser to permit unrestricted attacks on London, including such targets as government offices, banks, railway stations, etc. The indiscriminate nature of the recent British attack on Karlsruhe was cited as a reminder that such action was not unprecedented. These efforts were accompanied in 1915 by a growing military influence on the course of German politics. Eventually these unrelenting pressures forced the Kaiser to permit London to become a target. The stage was set for the Zeppelin raids of 1915. Although the Germans were destined to face difficulties of their own, including navigational problems, bad weather, and the effects of rivalry between the air arms of the Army and Navy, nonetheless they enjoyed a number of advantages, both strategic and physical. London, with its extensive docks and large ring of industries in and around the metropolitan area, was close to German bases.[12] Attempts by the English to black out coastal cities and towns, not to mention London itself, were ineffective.

Few social historians have noted that the "black-out," so closely associated with World War II, was a legacy from World War I, when lights were extinguished on London streets and dimmed all over the eastern counties to prevent enemy airships from finding targets. But unlike the emergency ordinances of the Second World War (referred to in the literature as A.R.P. [Air Raid Precautions]), no decision was ever reached during World War I to make the blackout total all over Britain. Such "dim-outs" as were ordered offered no direct defense of London, nor did the first efforts to provide a modicum of protection to civilian populations through the use of anti-aircraft guns or barrage balloons. At most, these combined efforts may have made the task of night bombing more difficult. Daylight raids were always an alternative; the Germans did not fail to use this option. From the outset they realized that so vast a city as London could not be hidden from the sky, by day or night.

Throughout the Great War the British were preoccupied with the land war, mainly on the Western Front. The clash of great armies was considered a necessary prelude to any eventual attack on a nation's capital. (As late as 1937, it was still the view of military experts that one of the first attempts made by a belligerent should be to strike a paralysing blow at the most vulnerable point in the opponent's territory, the administrative heart of the country. In 1939, it was thought that a few well-directed bombs aimed at Whitehall might tip the balance in favor of the Germans).

Thus, the British War Office spared little thought for home defense against the German airship raids of 1915-1916, recognizing perhaps that, because of their long range, a disproportionate effort would be required to provide a defense against Zeppelins over so large an area as London and the eastern countries. The first regulation giving civilian authorities power to control ground lights was contained in an Order in Council of August 12, 1914, under the authority of the Defense of the Realm Act, Clause 23, which empowered the competent naval or military authority at any defended harbor (that is, one declared to be "defended" by the Admiralty or Army Council) to issue an order directing that all lights, other than those not visible from the outside of the house, be kept extinguished. On September 1, 1914, this regulation was extended to include any "proclaimed area", and on September 17 the first general regulation was issued (D.R.R. 7A) authorizing the Secretary of State to order the extinction or dimming of any lights anywhere. This regulation was the occasion for the first protest by the General Staff against what they conceived to be an encroachment on their responsibilities. By decision of the Cabinet, the control of anti-aircraft guns and anti-aircraft defense generally, and in particular the defense of London, had already been vested in the Admiralty.

By that time the Commissioner of London Police had issued public notices requesting that all bright lights used for advertising purposes be dispensed with, and that the illumination of shop fronts be reduced. As a result of observations made from the air, it was found that these notices had been ineffective, so on October 1 they were reinforced by an Order, repeatedly renewed. Street lamps were extinguished, or shaded to break up all conspicuous groups of lights visible from the air. This order, extended one month at a time, by December 9 was further strengthened by a paragraph prohibiting the use of outside lights of all descriptions. All vehicle headlights were dimmed and had to carry a red rear light between one hour after sunset and one hour before sunrise. (Regulation 7A now became Regulation 11).

Attempts to impose similar regulations reducing all lighting outside the city of London were accompanied by a great deal of overlapping of authority and lack of coordination in enforcing anti-aircraft measures. Conflict soon developed between representatives of the Admiralty, which claimed responsibility for all inland places where there was a "garrison," and the War Office. The Admiralty by then had initiated a mobile anti-aircraft service without prior consultation with the War Office. Similarly, the War Office had issued various instructions concerning the reduction of lighting about which the Home Office had not been informed, whereas the Home Office's supposed responsibility for these arrangements rested with the Admiralty. These jurisdictional disputes were to plague the development of a rational home defense system throughout most of the first two years of the Great War.

The existing Grimsby Order, as it was then called, was replaced, at the request of local authorities, with a more stringent one requiring that all lights be extinguished or absolutely obscured from one hour after sunset until one hour before sunrise. (In the larger towns and cities it was necessary to maintain some lighting in streets and on railway premises, and in factories, where lights could not be obscured without interfering with essential war production).

In October, 1914, the Admiralty Office made tentative arrangements with chief constables within a radius of sixty miles of London to provide that the police would report by telephone to the Anti-Aircraft Section any enemy aircraft seen or heard overhead. In January, 1915, this arrangement was improved and extended to include the whole of East Anglia, Northants, Oxfordshire, Hampshire and the Isle of Wight. Unknown both to the Admiralty and the Home Office, it appears that at the instruction of a section of the War Office, all chief constables had been requested to make similar reports by telegram to the War Office. At some stage in this hopeless tangle of jurisdiction, the police were also instructed to report in all cases to the Admiralty, which would in turn pass on any information thus received to the War Office!

The Admiralty meanwhile had arranged to warn the Railway Executive Committee, who were in direct telephone conversation with the headquarters of the principal companies, as soon as they had any information that an attack was likely; they were to inform the Commissioner of Police at Scotland Yard if and when hostile aircraft had reached the country, and they were given even more ambiguous instructions to report if these movements pointed to "an intention to" attack London. But the Admiralty did not undertake to communicate warnings to the outlying localities, and the civilian authorities were left to make the best arrangements they could to obtain information from one another, or from naval or military authorities.[13]

After several German airship raids in 1916, rumors of the approach of hostile craft became more frequent. On a number of occasions (perhaps more than the authorities would like to admit), emergency measures were put into operation on instructions from the police or military authorities without any real justification. An urgent need resulted to place the responsibility for giving air raid warnings into the hands of a single authority, which could act only on definitive information that hostile aircraft had actually reached or were about to reach a given locality or district.

Factory managers had discretion over whether to extinguish lights at night; no general orders had so far been issued. In time, similar discretion was given to responsible railway and dock officials. Existing orders were modified during the summer of 1915 on points of detail, but no alterations were made in the general scheme of restricted lighting. After the January, 1915 raid on King's Lynn, it was seriously suggested that enemy pilots had been guided in pin-pointing their targets by the headlights of motor cars travelling on nearby roads.[14]

The problem of protecting London from the air was a peculiar one; conditions governing its defense found no parallel in any other part of the country. Anything approaching an attempt at total concealment of this huge metropolis was hopeless. Zeppelin commanders could base their calculations of position as much on the contours of the Thames estuary and the river itself as on any concentration of lights at night. The Thames formed an almost unmistakable guide to the heart of London, where the volume of street traffic after dark required at least some lighting. Rapid extinction of street lighting was an impossible task at the time, owing to the enormous number of gas lamps which could only be extinguished individually. The absence of street lighting would also increase the difficulty of dealing with fires and other damage resulting from air attacks. As a result of all these circumstances, the limited objective first sought was a uniformity of subdued lighting, in the hope that it would prevent the enemy from identifying specific objectives.[15]

Early in January, 1917, a conference was held to consider how the requirements of military traffic on dimly lighted streets could be met. Representatives from the military explained that recent observations from the air had shown conclusively that if any useful light were maintained in the streets, it would be impossible "to conceal the whereabouts of the metropolis from any aircraft approaching within twenty-five miles, or to conceal its topography from any observer passing over it." The main outlines and features of the city were clearly defined by the lighting along the main arteries, and by the contrast with unlighted spaces such as the Thames and the city's parks. Existing restrictions on street lighting, it was argued, were therefore not adequate for defense. While resumption of normal street lighting was considered to be out of the question, Army General Headquarters advocated increased diffused lighting, at least in London's central core. Details for such an arrangement were eventually settled upon between the Commissioner of Police and the engineering department of H.M. Office of Works, acting in concert with local authorities. Accordingly, a new order was issued on January 25, 1917.

No further alterations were made in the scheme for defensive reduction of lighting in London. It was also discovered that as a result of the economy measures taken in dimming street lights and eliminating advertisement lighting (Order of May 29, 1917), consumption of coal was reduced by at least one-third that summer.

The provisional orders that came into force in January, 1916 mark the end of the first period in the development of air raid precautions and a civilian warning system. The official position was that lights that could be visible from the air or at sea should be prohibited, subject only to necessary exceptions, on all parts of the coast. The eastern and southeastern counties were covered by a uniform code of regulations which provided for a very stringent reduction of lighting; similar conditions applied in a number of inland areas of industrial importance within the probable range of enemy activity.

General responsibility for defense measures, particularly as they applied to London, remained with the Admiralty, although the War Office claimed authority over all inland places where there was a "garrison." To the extent of the increased number of camps and military installations throughout the country, the army thus established the right to be consulted on all questions connected with anti-aircraft precautions.[16] Where intelligence matters were concerned, arrangements had been made whereby all reports of the movements of enemy aircraft were eventually centralized in the Admiralty Office, but, as has been seen, there was as yet no provision for warnings to be issued to civilian populations by any central authority.

On the night of 31 January-1 February, 1916, nine German airships, after making landfall on the Norfolk coast and the Wash, penetrated a considerable distance inland, some of them reaching as far as Derbyshire, Staffordshire and Worcestershire. A number of towns attacked (including Burton-on-Trent and Walsall) were not then covered by dim-out or black-out orders, and owing to a breakdown in local warning arrangements, warning was not received in time for precautions to be taken. This breakdown was due in part to the fact that in spite of instructions issued to chief constables a year earlier (May, 1915), adequate arrangements had not been worked out for all cases. It also resulted from congestion on telephone lines, overloaded by "members of the public" making "unnecessary calls." A notice was issued to the press by the Postmaster General a few days later appealing to the public to use the telephone as little as possible on such occasions. Police circuits were kept free for official messages only.[17]

The immediate result of the raid in January, 1916 was a demand, voiced mainly by the Lord Mayor of Birmingham, that the "lights-out" orders should be extended to areas where they had not been in force, and that some new system be devised whereby reliable warnings could be issued to important population centers throughout the country. "Lights-out" orders were by that time applied to the whole of England, for it was now evident that the Midlands and even western parts of the country could no longer be considered immune from attack; but an effective warning system would require time for full development. Nevertheless, the "lights-out" order would be given somewhat later further inland. There was a corresponding extension of the order dimming headlights on vehicles.

In Midland areas containing munitions factories, where an air raid observer system was in force, communications between areas were unreliable and telephone lines were congested. Elaborate precautions had been taken as far north as Nottingham, Bath, Gloucester and Worcester, but traffic was held up on the railway to Hull and the northeast coast for no adequae reason. The public was tense, and there was a risk that munitions workers would refuse to work the night shift unless they could be assured of warning in sufficient time to evacuate. This was no simple question of putting out the lights. It was found necessary to issue a circular to chief constables asking them to attempt to arrange to warn munitions factories in their areas by telephone or other system, such as whistles, hooters, sirens, etc.[18]

On February 10, 1916, the War Committee of the Cabinet met to work out a more effective policy. The Cabinet decided that the War Office, not the Admiralty, should assume responsibility for the defense of the country against air attacks, and that police reports should be communicated to General Headquarters, Home Forces. The country was divided into districts. At first warning these districts did not, as expected, coincide with police areas, the unit previously used for this purpose. Districting, as it was carried forward, was also governed by the existing arrangement of trunk telephone lines, and by military considerations.

Eventually Wales and parts of Scotland were considered liable to enemy attack. Seven "Warning Control" districts were set up, of which one comprised all of Scotland. The remaining districts covered England and Wales. Each was placed under the authority of a warning controller, who also served as Anti-aircraft Defense Commander. Headquarters for each warning controller were situated at a main center of the telephone system. District controllers were responsible for the collection and transmission of all information regarding the movement of hostile aircraft. Each "Control" was divided into numbered "Warning Districts" and the controller was responsible for warning the districts within his "Control." The boundaries were selected more or less arbitrarily. They were to be of convenient size and shape and could not be altered except in minor details.[19]

The most important source of information for the Warning Controllers was a system of observer posts established in various parts of the country. In the case of London, a special cordon was established around the city outside a radius of forty-five miles from the center to warn of attacks. In the case of important industrial centers, observers were placed at points about thirty miles distant from the boundaries of the districts, on the side most exposed to attack. In addition to these outside cordons, which were fifty to sixty miles from city limits, cross cordons were established to prevent enemy aircraft from moving parallel to and between the main cordons and coastal cordons. Wherever possible, observer posts were connected to Warning Control Headquarters by direct telephone lines. Each Warning Controller was also in communication by direct telephone with his Chief Warning Controller, various anti-aircraft stations, and adjoining Warning Controls. They also received reports from troops, naval installations, police, and railway officials within their areas.[20] This system was originally introduced when Field Marshal French commanded the Home Forces, and was kept for convenience sake after he relin-

quished his post. There were four warnings, issued in stereotyped form:
1. Field Marshal's Warning only
2. Field Marshal's Order: Take Air Raid Action
3. Field Marshal's Notice: All Clear
4. Field Marshal's Order: Resume normal conditions

No information was given as to the whereabouts, number or course of enemy aircraft. Exceptions were necessary for the south and southeast coast where attacks could and sometimes did impend before information could be communicated to the Controller and a warning issued. When, after a raid, Warning No. 4 was sounded, all emergency precautions were withdrawn.

The transfer of general responsibility for anti-aircraft precautions to the War Office was made on February 10, 1917. In January, 1918 special warning districts were abolished. Daytime warnings would be issued in the same way as at night. The effect was to localize the disturbances to morale caused by the issuance of warnings by confining them more closely to the area threatened. As a matter of record, enemy aircraft never penetrated far inland by day.

By this time the speed with which reports of approaching aircraft could be transmitted had greatly increased. For example, in a night raid during September, 1917, London Control Center received eleven different messages from the Essex Police in a half hour, each giving notification of a fresh development in the attack, and all these messages except one were received within three minutes of the observation. There was, to be sure, a certain amount of duplication, but the system was greatly improved in comparison to the haphazard attemps made in the first year of the war.

We now turn from the overall strategic view of air defense to the tactical, local view. Britain, practically undefended at the opening of the war, had seemed an easy target. At first the British had attempted, in an unorganized way, to protect their civilian population and defend the capital city, occasionally even finding themselves capable of destroying an enemy airship; but they soon learned that ground defense had to be organized. The sound warning system was sorely in need of streamlining. Lights were dimmed or extinguished on the streets and screened on the waterfront. All illumination for advertising purposes was forbidden. Windows were covered so that at night London became dark. The Zeppelins, compelled to fly at great heights by anti-aircraft guns, were blinded.

Observer posts were originally manned by military personnel unqualified for active duty. They were organized into fourteen companies of the Royal Defense Force. From the outset, discipline was poor. Personnel had to remain on duty from sunset to sunrise, whether a raid was anticipated or not.

Eventually military personnel were replaced by police, except on the east coast and in parts of the southeastern counties more liable to sudden air attack. Observation was continuous, day and night, except during bad weather. Eventually, over 200 observer posts were manned by the police, releasing the military personnel for other duties. Many observer posts in the County of London were connected by central telephone lines with the Central Observation Room at County Hall, Spring Gardens, manned by 1,200 civilian volunteers.

After February 24, 1917 new arrangements went into effect. Warnings would be issued as follows:

1. Preliminary. . .in form: "Home Forces Warning: Zeppelin crossed coast at _____(place) at _____(time) proceeding in _____ (direction)."
2. An order in form: "Home Forces Order: Take Air Raid Action."
3. A further order in form: "Home Forces Order: Resume normal conditions."[21]

The first order, to be issued in all Warning Districts, was intended to be confidential. The second was issued to Warning Districts only when threatened with an attack, as a signal for extinguishing any obscured lights and putting other emergency measures into operation. The third was issued when all danger was past, to all who had received Order No. 2.

In operation these arrangements were found to be defective on several counts. After preliminary warnings, it became impossible, for example, to prevent the issue of Order No. 2 from becoming known in munitions works, with resulting panic and disorganization during which valuable working time was lost. As in the past, there were false alarms. Doubts about the system were confirmed during raids which came in early March. By the time warnings reached all members on the General Warning lists the information was already often obsolete. Many factories had no lights to be extinguished and required no warning at all. Additional delays were caused by persons who failed to answer their telephones!

The upshot was the abolition of Order No. 1, the general preliminary warning, leaving three stages:

1. Field Marshal's Order: "Take Air Raid Action"
2. Field Marshal's Order: "Resume normal conditions," and
3. Field Marshal's Notice: "Country reported clear of hostile aircraft."[22]

As a result of drastic reductions in street lighting, pedestrians commonly used electric torches (flashlights) to find their way about after dark. If the use of torches became excessive, a warning was to be issued that they would be prohibited. The question even arose whether the striking of matches was to be permitted along the sea front—or elsewhere in the open. In a number of cases, proceedings were instituted against offenders and convictions secured, but the view of the Admiralty and General Headquarters, Home Forces was that such precautions were unnecessary; chief constables were so informed.[23]

Ambulance and fire brigades were organized, and other kinds of rescue to deal with casualties, fires and other damage caused by bombs or gunfire. Reliable warning was essential to the effectiveness of these arrangements. Civilian authorities were primarily dependent on information from military sources and acted on military advice. From the outset, "passive defense" was considered to be of great military importance.[24]

The first period in the development of an air defense system, terminating in January, 1916, was improvisational, almost experimental. It may be summarized as follows: anti-aircraft precautions required dim-outs, if not total black-outs, and consisted mainly in the reduction of lighting at night in London, the eastern and southeastern counties, and some of the more important industrial centers in the Midlands. During this period responsibility for anti-aircraft precautionary measures and enforcement rested with the Admiralty.

As has been shown, no serious attempt was made by the enemy to penetrate far inland during this early stage, until the night of January 31, 1916 when German airships penetrated far inland over the Midlands. Important changes in policy were made operative over the whole of England, Wales, and much of Scotland. The necessity of a more effective early warning system resulted in the transfer of responsibility to the War Office. The new warning system, when it came into full operation, made possible the expanded maintenance of lighting for essential purposes and reduced to a minimum the former degree of interference with important and even vital industries.

With increased enemy activity during daylight hours, and airplane attacks on London in the summer of 1917, the whole question of air raid warnings to the general public came under renewed scrutiny. Sound signals to give local warning to industrial establishments had been tried in a number of towns, and had resulted in all the inhabitants being made aware that an attack was imminent. In some cases this was welcomed by the public, but in others it had to be abandoned because of unfavorable public reaction after a short experimental period. Eventually, however, a uniform public warning system was set up in London, for daytime use in the first instance, and later at night as well, including provision for public air raid shelters to minimize casualties.

Modern readers will be startled, perhaps, by the quaint approach initially adopted by British officialdom early in the war. When the first warning of an approaching airship or plane came, local police authorities then awaited the Commissioner's "Take Cover" warning. Immediately upon receiving this, police officers went hurriedly through the streets, on foot or on bicycles, flashing "Take Cover" notices printed in red letters. Local police stations, authorized to send out sound signals, would fire two rockets. When the attack was known to be over, the "All Clear" signal would be sounded at the station and the police would again go out into the streets with "All Clear" placards. Regular and special constables, air raid relief workers, and ambulance details, as well as doctors and nurses called out for the emergency, were then dismissed unless their services were required to attend to casualties.

In his book, *The Defense of London, 1915-1918*, Captain Alfred Rawlinson, an anti-aircraft officer, admitted that there was no adequate defense against attacks from the air.[25] Before a system of ground defense was gradually worked out, more than 500 civilians had been killed, more than a thousand injured, mainly women and children, and over 2 million pounds worth of property destroyed. The combined efforts of the Admiralty, the War Office, the Home Office and local constabularies did not provide Londoners with an effective public warning system. None of the numerous alternatives proposed seemed more effective than "the maroons," which were a basis for the warning system later refined and retained throughout the remaining years of the war. The most important question faced by British authorities was whether (assuming that it was desirable at all to inform the public in large cities like London that an air attack was expected, and this was evidently taken for granted by all but a very small minority) it was possible to give a prompt and reliable warning. Not until the responsibility for air raid precaution was assumed by army authorities in 1916 was there a moderately (but never fully) effective air raid warning system.[26]

Even when such a system was put into effect, unforeseen difficulties arose. To begin with, it was by no means certain that the effect of the warning siren would be to clear the streets. During early raids, before their potential for danger was fully realized, public feeling at being regimented in this way was one of indignation, combined with curiosity rather than apprehension. Public officials faced the risk that general notification of an attack might attract large crowds of people, who under normal circumstances would have remained indoors, out into the streets. Then as now, people unmindful of their own danger were by some strange inner compulsion attracted to scenes of horror and destruction.

Formulators of new policies at first were lured to the tentative conclusion that sounding a general warning would have the effect of reducing casualties by enabling people assembled in theaters, concert halls and cinemas, for example, to disperse in time to avoid mass danger. Such conclusions were almost certainly illusory; planners soon found themselves forced to accept that a direct hit on a crowded theater would result in the loss of many lives, and consoled themselves by calculating that the chances of such a misfortune were slight. The risk, it was maintained, would be greater if the audience were given the opportunity to leave the building, which they would almost certainly take.

Crowd psychology in 1915 was still in its infancy. Honest if untrained official minds wrestled with conundrums. Unless warnings were sounded in ample time, crowds would be caught in the streets during an attack. If, on the other hand, warnings were given in ample time, crowds would still be unable to disperse very far, and the larger the crowd, the greater the problem of dispersal. Traffic would be interrupted, congestion would develop, and air raid shelters might provide no safe haven. The risks involved in the development of an early warning system were not yet fully understood, if indeed they ever would be. The general public was by no means unanimous in favoring such an arrangement as was gradually evolving. The British people above all seem to have cherished their freedom of mobility, even in the face of danger from the skies.

In the summer of 1916 authorities in each district were left free to make the best arrangements they could for obtaining information about air attacks and distributing it to the surrounding populace. Experience soon revealed that such arrangements were ineffective. There was still no guarantee that hostile aircraft might not approach unobserved or unreported, which would prove fatal to any sense of security based on a promise of public warning.

The authorities were trying to avoid the previous chaotic actions taken on the basis of uncorroborated reports, resulting in false warnings issued on numerous occasions when no attack materialized. At Hull, for instance, before October, 1915 sirens were sounded on nineteen different occasions, on only three of which bombs were dropped in the neighborhood; at Ipswich, before February, 1916, police received 44 warnings of approaching enemy aircraft, but aircraft passed over the town only six times and no bombs were dropped.

If a public warning system had been in effect in London during 1915 and 1916 it would have been necessary to issue warnings on 64 occasions during each of these years, while the number of attacks that actually materialized was five. Technical defects in the revised system could be and were eventually remedied, but the majority of chief constables who had had any experience with public warnings concluded that they were unnecessary and undesirable. The general effect was to bring crowds into the streets, causing considerable alarm and excitement where none was warranted. As a result, after the initial trial period in a number of towns the revised system was abandoned. In others it was continued only because the populace had become accustomed to it.[27] Eventually the War Committee of the Cabinet approved the system of issuing public warnings, not as a universal measure, but only where local conditions required it.[28]

In general, the principal employers of workers during the war were almost unanimous in their opposition to general alerts or public air raid warnings. They felt that the proper course for them to follow was to maintain "perfect silence and . . . perfect darkness." In Birmingham, it was demonstrated that 10,000 to 15,000 workers could quit the premises in less than 15 minutes with no apparent disorganization or panic.[29] Reliance on transmitting orders to initiate sound signals also was delivered a heavy blow in March, 1916, when a blizzard damaged telephone lines.

The problem of London remained a special one. The sound of exploding bombs was inaudible beyond a few miles from the scene. Disturbances resulting from Zeppelin or airplane attacks were small, localized affairs. Any warning system, to be effective, would have to be extended over the entire, vast metropolis. It was anticipated that the disadvantages of the proposed warning system, as experienced in the smaller provincial towns, would only have been intensified in London, where the difficulties of issuing such warnings, or of controlling large populations, were far greater. Not until 1917, when conditions had materially altered, was it finally decided that a system of public warning for the London metropolitan area was practicable and desirable.

It was the alteration in the nature of air attacks which ultimately forced a change in policy. The German High Command, having realized by the end of 1916 that the expense involved in the manufacture and operation of Zeppelins was not justified by the damage they were able to inflict, decided to launch airplane attacks on England. A gradual change in British policy was also brought about, including the education of the public to the greater risks to which they would be exposed during air attack and the steps that could be taken to avoid them. A factor influencing the new policy was the introduction of the Home Forces Warning System, which made it possible to notify the public of the approach of hostile aircraft with increasing rapidity and certainty.

From past experience it was concluded that no one course would be suitable for all occasions, but, in general, the public was advised that persons caught out in the open in an attack should seek cover, and those already indoors should remain there. Notices to this effect were issued by the Commissioner of Police. Similar warnings were embodied in posters illustrating the various types of British and German aircraft being used in combat and on bombing missions. These posters were distributed in large numbers and achieved a considerable popularity. Advice was also made available as to the best means of dealing with fires caused by incendiary bombs. A series of useful warning leaflets was prepared and distributed by the British Fire Prevention Committee, an unofficial body of experts who made a special study of fire prevention and precautions.

On June 13, 1917, shortly before noon an attack was made on London by a squadron of fourteen Gotha airplanes, causing considerable damage and resulting in 162 persons killed and 426 injured, including many casualties from falling fragments of anti-aircraft shells. Estimated property damage from this raid was placed at 125,000 pounds. This was the third in a series of daylight raids by airplanes, following two on May 25 and June 5, marking the beginning of a new phase in the German air assaults. Previous raids had taken the form of either night attacks by Zeppelins, or daylight attacks by single airplanes or seaplanes. (On a few occasions groups of planes reached the coast, but made off almost immediately afterward). Henceforth most attacks on the countryside, and all of those on London (with a single exception), were carried out by heavier-than-air craft, at first by day and later by night.

This turn in the nature of the air war introduced new and unprecedented factors. The speed of airplanes was much greater than that of airships, giving less warning time and greater surprise. Daylight bombing occurred when city streets were crowded. The result would be heavier casualties, unless advanced warning was received in time for civilians to take cover. More accurate observation of approaching hostile aircraft was possible during daylight hours, now more favored for attack, but this advantage was offset by the speed of their approach. Interruptions of work were greater, and rapidly becoming intolerable. In the light of these new developments, the urgent necessity for reintroducing a uniform public warning system in and outside London had to be considered.

The first step taken was to consult the civic authorities in the metropolis, followed by a conference of mayors on June 21, 1916. The consensus expressed at these meetings was that public warning was desirable and necessary, with the caution that such warnings had to be accurate. The Cabinet, however, in a meeting on June 26, decided against introducing a public warning system. Three more raids followed on July 4, 7, and 22, the first and last of which did not penetrate beyond the Essex coast. But on July 7 German raiders over London caused serious losses, with 53 killed and 182 injured in east central districts of the city, such as Stepney, Shoreditch and

Stoke Newington.

The subject of a public warning system was reopened. On July 10 the Cabinet finally gave approval to new arrangements, already under way in some locales, to warn Londoners of enemy air attack. It was decided to give a five minute advance warning within a radius of ten miles from Charing Cross, which would cover the main populated districts of London, allowing 4½ minutes for transmission. When information was received that the enemy was crossing a line about 22 miles from Charing Cross it was immediately incorporated into the warning. The sources were a line of observers stationed at that distance from London.[30]

There was a flood of suggestions as to the means to be employed in sounding these warnings. They included ringing of church and tram bells, turning on street lamps, firing blank charges from artillery or anti-aircraft guns, and the sounding of "syrens." To all except the last there were fatal objections. All the others would either take too long or would be ineffective. Originally sirens were designed to be audible over a long distance at sea. It was now argued that the sounding of "syrens" would be drowned out by the overpowering noise of London traffic, unless large numbers of them with powerful volume were used.

Another suggestion was the use of rockets, the firing of which would be accompanied by clouds of colored smoke. Rockets had also been originally designed and used for firing distress signals at sea. Now proposed were sound bombs, or "maroons" fired from small brass mortar guns, bursting at a height of 1,000 feet. The proposal called for firing three such signals from police stations in the London Metropolitan area when an attack was thought to be certain; a plan for their use had been outlined in a letter to the Home Office as early as October, 1915.[31] The detonation of these devices was quite powerful, and this was first considered a main drawback to the use of maroons. It was thought that the sound would be mistaken for enemy bombs or gunfire. These objections were overridden by the obvious advantages to be gained from employing this swift means of warning. When the maroons were at first adopted for night warning signals in London they were slightly modified to show colored stars upon bursting. This was followed by the elimination of sirens as a means of giving warning in the metropolis, although they would be employed again before the war's end.

In the initial trial period two maroons were fired from each police station. Soon there were demands to accompany these signals with sound signals, especially at the onset of darkness, for by autumn of 1917 enemy airplanes had commenced night attacks on the city. "Take Cover" warnings were still being disseminated by constables on foot, on bicycles, or from motor cars, carrying the same kind of placards as had been used at the outbreak of the war. The only difference was that the later placards were illuminated. Nightly, constables circulated through the city streets, sounding their whistles, bells, and horns and calling out to the populace to "Take Cover." The authorities considered this a reasonably satisfactory way of clearing the streets, however quaint or primitive it may seem in retrospect. Repeated raids during the winter months required the adoption of a more uniform system. Maroons were finally adopted, despite the disturbance (especially at night) of their loud explosions. London public opinion was becoming more and more emphatic in its demand for something more arresting than constables on foot or wheels circulating through the streets displaying "Take Cover" signs. Meanwhile public authorities were busy establishing warning systems of their own, including more modern types of sirens, and colored lights. Finally, in January, 1918 the London Police Commissioner issued a notice that maroons would be fired until 11 p.m. Extension of their use to a regular 24-hour basis was only a matter of time. Upon the Commissioner's advice to the Home Secretary, it was decided that maroons should be used at any hour of the day or night. At the same time, the decision was reached to issue "Take Cover" notices at an earlier stage in the attack.

These various forms of air raid precautions were eventually extended beyond London to the entire southeast whenever enemy aircraft approached the English coastline. Sometimes such actions created unnecessary disturbances. As a refinement, warnings were ultimately issued by successive districts, rather than relaying the same notice from what was termed "The Large Day Warning District." At this final stage in the war sirens were not used to sound the "All Clear" signal; reliance was placed on bugles. The Police Commissioner was instructed to increase the number of buglers to make the signal effective.[32]

Despite the slow, faltering way in which the air raid precaution system evolved, eventually these arrangements provided London with a workable if not entirely effective public warning scheme, the best obtainable solution to a problem for which no perfect solution existed. This at least represents the official view of the Air Ministry during the period when its several Memoranda on the subject were being drafted.[33] As the experiment unfolded, and as the means employed changed from one system to another with little or no rationale, the disadvantages often outweighed the advantages, and vacillation resulted. Each successive system was freely criticized. Over the long run, none of the numerous alternatives or variations met the need for a distinctive signal more effectively than did the maroons. They formed the basis of the revamped warning system which was slowly and haltingly worked out during the first two years of the struggle and was retained throughout the last two years of the war until the Armistice was signed in November, 1918.

Notes

1. The definitive source for his ideas is Giulio Douhet, *The Command of the Air* (trans. by Il dominio dell'aria, New York, 1972, reprint 1942 edition). For a recent view see Aaron Norman, *The Great Air War* (London, 1968), 342-5.
2. The balloonist, Thaddeus Lowe, as a young man impressed high-ranking Union Army commanders. After he had made some spectacular ascents and sent President Lincoln the first telegram ever transmitted from the atmosphere, Secretary of War Cameron made him head of the new aeronautics section of the Union Army. Although the heavily wooded nature of the country and the difficulty of transporting and flying balloons limited their usefulness, Lowe sometimes obtained information of value to the Union commanders. For an authoritative account of this almost forgotten aspect of the Civil War, see Frederick S. Haydon, *Aeronautics in the Union and Confederate Armies, with a Survey of Military Aeronautics prior to 1861* (Baltimore, 1941); also Eugene B. Block, *Above the Civil War: the Story of Thaddeus Lowe, Balloonist, Inventor, Railway Builder* (Berkeley, 1966).
3. On Zeppelin, see Margaret Goldsmith, *Zeppelin: a Biography* (New York, 1931), and Hans G. Knausel, *Zeppelin and the United States of America* (Friedrichshafen, 1976), 7-10. For an account of his balloon ascent over St. Paul, see John Koster, "Count von Zeppelin's Visit to the Front," *Civil War Times Illustrated*, Vol. XVII, No. 9 (January, 1979), 12-17, and quotation, 17.
4. The Air Historical Branch was formed in 1918, shortly after the creation of the Royal Air Force, to prepare narratives for the Official Air Historian. This function was transferred in 1918 to the Air Ministry. In addition to preparing the official history of the air war in the War of 1914-1918, a short history was also produced for official use, along with squadron histories. With the completion in 1935 of H.A. Jones' and Sir Walter Raleigh's volume, *War in the Air*, the Operations and Technical sections became redundant and the Records section was absorbed as part of the Air Ministry Library. The Air Historical Branch was re-established in 1941 under the authority of the Air Council and its duties were extended to include the compilation of operations narratives. The Air Historical Branch Records contain papers from the period 1914-1918 received from the Air Ministry Departments, the RFC (Royal Flying Corps), and RAF (Royal Air Force) formations, as well as from other sources. Also included are registered files which originated in the Admiralty, War Office and Air Ministry. Not all papers referred to in the bound volumes I to VII (which form an alphabetical index to subjects covered in the collection) have survived.
5. Norman, *The Great Air War*, 342-5.
6. Barbara Tuchman, *The Guns of August* (New York, 1962), 175-6, 346. The Germans were obsessed throughout August with the goal of "intimidating the Belgians" into giving up "their stupid and futile resistance." Under a flag of truce, the former German military attache at Brussels, known personally to the Belgian Commander at Liege, General Leman, was sent to persuade him or, failing in that, threaten him to surrender. Leman was told by the emissary that Zeppelins would destroy Liege if he refused to allow the Germans to pass through. As the Germans overran Belgium and much of the Channel coast, the danger of air attacks on Great Britain became more serious.
7. For the fate of these three ships, see Norman, *The Great Air War*, 395-6; also H.A. Jones and Sir Walter Raleigh, *History of the Great War: The War in the Air* (6 vols., London, 1922-1937), I, 88.
8. Norman, *The Great Air War*, 350.

9. In Bachmann's view, successful raids on the enemy capital, in view of the nervousness of the British people, would pave the shortest way to crushing Britain. In his memorandum, Pohl argued that London was a defended city according to the Hague Convention's definition, and bristled with military targets. He promised the Kaiser that German airships would take every precaution to avoid damaging historical sites and private property. See Francis K. Mason, *Battle over Britain* (London, 1969), 17-18. Cf. Norman, *The Great Air War*, 348-52. On the internal struggle and Germany's over-arching war aims, see Fritz Fischer, *Germany's Aims in the First World War* (New York, 1967), 280-2.
10. Reichenau to German Foreign Secretary Arthur Zimmermann, August 25, 1914, quoted in Fischer, *Germany's Aims in the First World War*, 281.
11. Mason, *Battle Over Britain*, 17-18.
12. A map attached to NASF 170 indicates the bases in Holland from which Zeppelins could be launched against England. Air I, File 568, 16/15/126. As the Germans successfully conquered the Belgian fortifications at Liege and Namur and captured one Channel port after another, on September 3 Kitchener asked the then Secretary of the Admiralty, Winston Churchill, in Cabinet whether the Admiralty would accept responsibility for the air defense of Britain, as the War Office had no means of discharging it. Prior to assuming this responsibility, the Secretary wrote the first of several memoranda, dated September 1, 1914, to the Director of Air Division, Chief of Staff, in which he directed that the largest available force of naval airplanes be stationed at Calais or Dunkirk. He had received reports, he wrote, that the Germans would attempt to attack London and other places with Zeppelin airships, of which it was said a considerable number existed. He concluded: "The proper defense is a thorough and continual search of the country for 70 to 100 miles inland with a view to marking down any temporary airship bases, or airships replenishing before starting to attack. Should such airships be located they should be immediately attacked. Commander Samson, with Major Gerrard as second in command, will be entrusted with this duty; and the Director of Air Division will take all steps to supply them with the necessary pilots, aeroplanes and equipment." This memo is included in the collection of documents edited by Eugene Emme, *The Impact of Air Power: National Security and World Politics* (New York, 1959), 27. Churchill printed it in his *The World Crisis, 1911-1914* (New York, 1928), pp. 327-44.
13. "The Development of Lighting Restrictions and Warning Arrangements..." Extracts from Part II, III, and IV, prepared by the Air Ministry (February, 1916), with all supplements. Air I, 721, 46/4, par. 1-30, pp. 1-7.
14. On the question of motor lights, see letters to *The Times* (London), January 22, 27, and February 11, 1915, including one from Holcombe Ingleby, M.P. from King's Lynn, scene of one of the early attacks.
15. Air I, 721, 46/4, pp. 7-13.
16. *Ibid.*, pp. 13-15.
17. *Ibid.*, par. 54, p. 15.
18. *Ibid.*, pp. 16 ff.
19. *Ibid.*, par. 61, pp. 17-18.
20. *Ibid.*, par. 62-70, pp. 18-20.
21. *Ibid.*, par. 71, p. 23.
22. *Ibid.*, par. 71 ff., pp. 25-40. There were also, of course, many changes of detail.
23. Tables of local sunset times were issued during this period by H.M. Home Office, and after November of 1916 they went on sale in branches of H.M. Stationery Office in order to afford the public a means of determining the hour at which the various orders took effect in parts of the country. These tables were worked out and updated by the Astronomer Royal.

24. Air I, 721, 46/6, pp. 1-2.
25. *Defense of London, 1915-1918* (London, 1923).
26. Air I, 46/6, par. 167, entitled "The Experience of Provincial Districts."
27. Introduction of the New Warning System, 1916, in Air I, 271886/133.
28. Two regulations for this purpose were introduced under the Defense of the Realm Act, March 30, 1916. No. 25A and 25B were included in an Order in Council. Some areas retained sound signals, as opposed to the more widely used maroons or rockets, while in other locales sound signals were rejected. The effort at uniformity faltered.
29. Air I, 271886/338-341.
30. Air I, 46/6, par. 188, and 271886/509.
31. Air I, 271886/113.
32. Air I, 271886/779.
33. The remaining sections of the Air Ministry's lengthy Memoranda for these years deal with problems of the first two years of the War so that the experience would be available in the event of future wars. "It may be that in the event of another war," the Memo concludes, "these precautions, or the machinery on which they depended, may be impracticable or obsolete; but assuming that the main factors remain substantially unchanged, the organization as it existed at the end of 1918 may at any rate form the starting point from which a scheme to meet future needs may be elaborated." Air I, 721, 46/6, pp. 1-4.

PERCEPTIONS OF MILITARY HISTORY

The Challenge of the Eighties: World War II from a New Perspective— The Hong Kong Case

Kenneth Taylor
University of Alberta

Official histories are by their nature more celebratory and heroic than definitive. Official history is to history as Victory carillons are to the art of campanology. Yet, in Canada at least, it seems that academic historians consider the history of the Canadian experience in World War II to have been written in sacred codas which must be left to gather dust in respectful silence. Quite the contrary; they must be re-examined and amplified. The Official Histories in which the Canadian military experience during World War II is recorded were compiled by those who had contributed to the conduct of a just, heroic and victorious war effort recently concluded. The errors and defeats were to be briefly though fairly examined only insofar as they still reverberated in the public memory and could not be ignored. The thornier details could await a more propitious occasion. This attitude was reinforced by the atmosphere of the period in which the official historians worked, one in which military danger again apparently threatened the Western and Imperial alliance whose most recent history they were recounting. It was not a unique situation. The biography of Dr. Thimme, the principal editor of the German Official History of the First World War in the 1920s, shows, not that pressures were brought to bear on him to present the facts in a certain way, but how the attitudes which he and his colleagues brought to the task in the period in which they worked dictated the form and content of the finished product. In the parlance of today, in both cases, the historical account was pasteurized and homogenized. But war does not lend itself to such a process without distortion. It is not that tidy or simple. Nor could a defeat such as at Hong Kong in December 1941 be contained in the few dozen pages assigned to it in the British and Canadian Official Histories. The concerns to "tidy it up" were also reinforced by the spatial limitations imposed by the publishers, the respective government Stationery Offices. But in the case of the Hong Kong affair that process was taken a step further, and it represents a prime example of the inadvisability of continuing to accept the Official

Histories as the definitive works they purported to be.

In the interests of protecting the Atlantic Alliance and reducing any chance of friction between British and Canadian troops, particularly if the latter should come under the higher command of the former as seemed likely in the Cold War period, the historical record was tampered with by the then respective Chiefs of the Canadian and Imperial General Staffs, Foulkes and Montgomery. Even the very evidence which the Official Historians were to use was "edited" before they began *their* tidying up process. General Foulkes reported in 1948 to his Minister in a Top Secret Memorandum, "after discussing this whole question with Field Marshal Montgomery he [Montgomery] agreed to have these offending paragraphs taken out" of the Official report on the Defense of Hong Kong by the General Officer commanding that defense.[1] It was thought best not to remind their successors in the North Atlantic Alliance, by publishing an official record which must contain reference to it, that it was only a few years earlier in the history of the alliance that the poor morale and fighting efficiency of the Canadian troops at Hong Kong were, in Foulkes words, "interpreted by their British superiors as a lack of courage, willingness to fight and even in some cases cowardice," while "On the other hand this had caused in the minds of Canadian troops bitterness, lack of confidence and resentment in their British Superiors."[2] In the circumstances of that time, the danger of creating friction between the forces or their governments and High Commands was considerable. But the existing official historical record thus has been sanitized twice.

Even after removing the most spiky of the bones that might stick in national throats, the reaction of those thought to be qualified critics of the accounts in the draft Official Histories was not favorable, and later even the twice revised record was not to escape further revision at their behest. There are at least two examples in the case of the draft of the British Official History. Colonel Price, an extraordinarily intelligent, well educated, gallant man of wide experience and unchallenged integrity who had commanded one of the Canadian battalions in the defense of Hong Kong, complained bitterly to the official historians that the draft of the British account was "written in such a manner as to create a wrong impression as to intent and motive [in the course of the battle]" on the part of the Canadian Expeditionary Force. Contrary to the impression given by the British Official History, "There *were* plenty of Canadian Officers who had had battle experience in the first war and who were competent to judge as to the possibility of a successful outcome of the defense of the island;"[3] the draft history "casts a reflection on . . . senior Canadian officers which I greatly resent and about which I protested [in the prison camps] before to General Maltby. . ."[4] In his opinion, the report on the behaviour of the senior Canadian officers by the Brigadier commanding them "is not to be relied upon.

He was in a state of great nervous excitement and I believe his mental state was such that he was incapable of collected judgement or of efficient leadership."[5]

The reference to the protest to Maltby in the prison camp evidences an extraordinary situation of which the *existing* accounts of the Hong Kong Affair take no account. Research shows that immediately upon capture the various officers responsible for the major decisions taken during the campaign "rushed into print", metaphorically speaking, on whatever paper they could salvage. This was at a time when the confusion was so great and the Japanese control of the prisoners so loose that escape was fairly simple and several attemps succeeded. It is a strange picture, these senior officers busily scribbling away in their school exercise books amidst the wreckage of defeat, ignoring the opportunity to organize their men to attempt to reach freedom and fight again. But this scene and the desperate lengths to which officers went to preserve their accounts in the prison camps indicate what was at stake.

It seems, from reading between the lines in these stained and tattered documents, that from the time of the loss of the Shingmun redoubt, the ostensible key to the forward defenses, and particularly the abandonment of the mainland, it was apparent to the commanders of each of the national contingents that there would be, after the war (their faith in ultimate victory is also remarkable), a reckoning; the nationalities involved, Britain, Indian, and Canadian, using as counters in this contest the units which each had contributed, would blame each other for the failure to meet the pre-war expectation of a 90 day defensive battle while awaiting relief from the sea. In the British view, the senior Canadian officer always regarded his first responsibility to be not to his superiors in the Hong Kong garrison, but, in the words of the General Officer Commanding in a secret report to the British authorities, "to the Canadian Chief of the General Staff, to the Dominion Government, to the Canadian People." Thus, by implication his first responsibility was to preserve the Canadian units intact rather than suffer heavy losses defending an indefensible position.[6] Brigadier Price's strictures over the extracts from the official dispatch of the commanding officer [Maltby] were accepted and the final version of the British Official History was modified accordingly, despite Price's personal involvement in the outcome of the conflicting versions. Rightly or wrongly, the Official History had been sanitized once again at the behest of another party with a personal stake in the Official Historical record.

The reaction of Augustus Muir, the pseudonymous author of the regimental history of the Royal Scots, to the British Official History as originally drafted, was similar. In a 64-page memorandum supplied to the Cabinet Office, supported by sworn statements by survivors and by prewar aerial photographs of the mainland defneses (which the Official Historians had not known existed), he pointed out the minor errors which could be corrected easily, and which are only to be expected. But he then went on to offer what was almost a legal brief in defense of the Royal Scots' performance. Muir compared the expected losses due to malaria to the actual ones, and studied the withdrawal over ground the units had never before traversed and the foolishness of the Brigadier's failure to appreciate the Shingmun redoubt as the key to the defense of the mainland. (He overlooked the fact that it was constructed on a bluff overlooking the colony's main water reservoir, so that its loss also meant the loss of that irreplaceable facility.) He points out that the pillboxes in the defense line were poorly constructed; that the routes used by the Japanese in the attack had not even been considered in the pre-invasion calculations; and that the redoubt at night, when indeed it was attacked, was by orders manned by three solitary men. In Muir's words, it was not "impregnable," merely a "death trap." Muir went on to point out the reasons why he considered the defense weakened by Garrison H.Q. in allowing the "vested interests of civilians—European and Chinese—to take priority over military considerations." He attacked the conclusion that the capture of the Shingmun redoubt "altered the whole course of the campaign," since the Japanese had in any case cut it off due to the failure of the British High Command to use its artillery to break up the Japanese commander's movement to outflank the redoubt. He emphasizes that the Brigadier and the General Officer Commanding projected a defense of the mainland, *after* its fall, longer than that projected *before* the fall of the redoubt.

And so it goes on. Muir shows that the Official Historians, without the oral testimony on oath given to the Royal Scots Regimental enquiries, and without the usual documentary evidence, had no idea what the defensive positions were like in the battle for Hong Kong, or what had happened in them. They had taken the opinions of senior officers who had very poor control and communications and who had not gone forward to see what was going on, for reasons which remain obscure. The official opinions, as he said, were "not based upon a full knowledge of the facts."[7]

Interviews with survivors has shown that, in this at least, Muir was quite right. The Official Historians were on shaky ground in relying on the opinions and reports of senior officers in their account of what was a series of small tactical actions, a "Soldiers Battle." In consequence of Muir's statements the Draft Official History again was modified. In a sense it had been sanitized once again in response to Muir's partisan regimental views. But equally important is the implication that the Official Historians, working against time to cover an enormous amount of ground, were not able, even with the unprecedented resources placed at their disposal, to give more than a cursory glance at the mass of evidence available to them from the participants in a confused affair of myriad small actions. The case of Hong Kong appeared to be a minor affair in scale and long term effects; but this is a false appearance, a fact to which this analysis must return.

Omissions and errors in the historical record of World War II have been corrected since the official histories by hundreds of published works. These memoirs, campaign histories, and accounts of particular engagements serve as supplements to, commentaries on, and thus a corrective to the official histories. This is indeed a "growth industry," particularly where controversy is a natural concomitant. Nor is the war in the Pacific, or the Canadian military experience, exempt from this process. A historiographical essay could be written on the works dealing with Singapore, for example. The chief Public Relations Officer of the Canadian Department of Veterans Affairs told James Leasor "he had received a number of requests for help from other writers who had reconstructed the story of Dieppe Raid over the years and he personally wondered what new material could now be found..."[8]

Surely the pattern will prove the same in the case of the fall of Hong Kong. The record of such an important episode in Canadian History, one might assume, has been corrected. It was the first occasion in which large numbers of Canadian troops were employed in action in World War II, the only time when sizeable ground force was contributed by Canada to combat in the Pacific, the first important defeat in Canadian military history, with an entire brigade annihilated, the first occasion on which large numbers of Canadian troops went into captivity, and the first since the Boer War in which Canada contributed to the active defense of the British Empire as such.[9] Surely monographs, scholarly articles, and certainly popularized versions of the battle and ensuing imprisonment would have poured from the presses. In this conclusion one would be wrong.

The popular press and the magazine circuit have written up the affair about once a decade on average. There exists one Masters Thesis on the role of the Royal Rifles; it was written under the patronage of the chancellor of the student's university, who happened to be the former commander of the regiment. According to the author, not a single request has been received for it from academic circles. There have been other attempts to write on the subject. At least ten manuscripts are languishing in dead men's attics and several more are under way by the living; but they fall into the "stand by your memoirs" category of dusty military recollections. Proof of the lack of interest in the defeat at Hong Kong on the part of the military historians in Canada and Britain, and their willingness to allow the official histories' accounts to stand without reexamination, is that when this research began a couple of years ago, many of the germane British and Canadian documents had to be declassified. No one had asked to see them since the official historians had completed their work.

The British publication picture is even more odd, despite Hong Kong's importance in Imperial history. In the immediate aftermath of liberation, several accounts of prisoner of war experiences and escapes by veterans of the Hong Kong affair appeared, as did similar accounts by those captured at Singapore or workers on the Death railway. They tended to be mass market paperbacks of poor quality, with the usual sensationalized titles and suitably exotic covers. There has always been a ready market for such material, and the Death Railway and Singapore examples still sell briskly in North America and Britain; but for Hong Kong examples one has to be satisfied with brittle and ancient copies from dusty second hand bookshops.

There have been two equally odd recent developments. Mr. Ford, who served with the Royal Scots in Hong Kong (together with his brother, executed by the Japanese for his role in the pipeline out of the camps to the outside world and decorated posthumously for his refusal to betray his comrades), felt that he had something to say, both about the defense of Hong Kong and his role in it, and the imprisonment which followed its failure. From such an extremely able and literate observer, with a trained legal mind, now serving as a senior civil servant in Scotaland, one could have expected much. But to preserve his anonymity as one of the leading figures in the United Kingdom legal system, he saw fit to fictionalize his recollections and views. Consequently, both of Mr. Ford's books are difficult to use as historical evidence.[10]

There is the case of Tim Carew, a veteran British officer in the war against the Japanese, decorated, wounded, and a first class author of popular military history. Carew turned his attention to Hong Kong as an unfilled gap in British Military history, committed to what he saw as the telling of a neglected tale.[11] His first book on Hong Kong was one of the earliest works in Pan's British Battles Series, coming out in 1963; but it went out of print and has not been reinstated in the original series. There is also the account of a major in the Grenadier Guards, once stationed in Hong Kong where his interest in the battle germinated. Impressed, after walking over the battleground, with the heroism of those charged with such a difficult task, and having met many of the civilians who had been involved in the siege and were imprisoned in its aftermath, he felt that their story was worth telling. The facts he learned, particularly from the civilian participants, add to the possibility that a professional military historian will be able to reexamine the defense and fall of Hong Kong in full, including the diplomatic and international ramifications and the Canadian constitutional and legal issues. But the appearance of his book in 1978 created very little stir, despite its novelty as the first serious work on the battle since the official histories, and it is out of print already. It left too many questions unanswered, since it hewed to the official British line: that the commanders were exemplary but the defense plan was unworkable, and that the regular Imperial and Indian troops were generally up to the best traditions of the British and Indian armies, whilst the Canadians performed inadequately. This was hardly the case.[12]

There are regimental histories, but they often suffer from the same difficulties as the official histories. Yet, what is left out of the regimental histories is revealing, particularly from the Canadian point of view. Nothing is said about the prompt convening of a regimental court of enquiry into the loss of the Shingmun redoubt in the prisoner of war camps, nor of the postwar unofficial enquiry in the Edinburgh Royal Scots Club, before which each officer who survived the camps was summoned on his return to explain his personal behavior in the battle five years before. The proceedings of this enquiry were taken down by a shorthand reporter, sealed, and deposited in the care of the Princess Royal, then Colonel-in-Chief of the Regiment. One finds nothing about the communications failure between the prepared positions in the mainland defenses, largely due to the fact that poor rural Chinese dug up many of the copper telephone cables to use or sell as fast as they were laid. Nothing is said about the intelligence officer for the mainland brigade who had been released from hospital 48 hours before the invasion and was walking the defensive positions for the first time to fix them in his mind when the Japanese attacked. There is no reference to the fact that many of the dour and experienced Lowland Scots of the prewar battalion had been returned to their homeland to flesh out the new battalions being raised there, and had been replaced by drafts of Welsh and North Country men with little training, no experience, and little in common with the Scots alongside whome they served. More important, only 1 in 10 of the trained regular NCOs and officers remained to direct the defense in 1941. The balance was made up of newly promoted NCOs, NCOs from Royal Scots Territorial battalions who had joined for the extra money during the Depression, and officers who appreciated the social and professional advantages in the Edinburgh of the 1930s of service in the Regiment. The situation became so desperate just before the Japanese attack that men who had been commissioned into the H.K.V.D.C (the Home Guard of the Colony) because of their standing in the business community were transferred into the Royal Scots to fill empty command postings.

There are no regimental histories of the 5/7 Rajputs and 2/14 Punjabis, since the passing of the British Raj so shortly after the war made such an eventuality unlikely. There is an Indian "Official History," and it contributes modestly to a general knowledge of the battle with its short, tactical, and heroic overview.[13] But regimental magazines did exist until 1948, and these are far more useful. A careful study of the postings to and from the battalions in the period 1940-1941 confirms one of Carew's opinions: "The Indians were a mixture of [recently arrived] reservists—time expired men recalled for duty at the outbreak of war and raw youngsters. The Rajputs and Punjabis, martial races both, have produced some of India's finest fighting men. But a reservist of any nationality, his service behind him, civilian life in front of him, must inevitably be something of an 'unwilling warrior.'"[14]

Carew had missed two additional factors further reducing the effectiveness of these particular battalions. As a result of the recruiting situation in India at the time, the traditional policy of keeping like with like among the quarrelsome tribes, castes and religions of India had been abandoned in certain cases. Where fear of immediate action was least pressing, for example in Hong Kong, the old rules, based on decades of experience and good sense, were suspended. Consequently the 5/7 Rajputs contained fewer of that outstanding Hindu "martial race" than it should have and the numbers had been made up with their avowed rivals, if not outright enemies, Muslims from the Punjab. It was unlikely that they could be welded into a cohesive fighting unit. Also, the battalion included Indian officers instead of British, as this had been one of the few Indian Army regiments chosen for the experiment of commissioning through an Indian equivalent of the Royal Military Academy (Sandhurst). The new policy was in its early stages, and as the Indian officers had very little in common with their men, coming as they usually did from entirely different cultural backgrounds, it was to prove a weakness in the forge of combat. The 2/14 Punjabis were homogeneous in the sense that they were all Muslims, but they were broken down into platoons made up of distinct and separate tribal groups within the same companies. Furthermore, both battalions had been brought up to British officer strength by introducing strangers from the reserve into the ranks, a weakening of that very trust by which the traditional Indian Army regular battalions lived and fought. And these two battalions of fine regiments with long and honorable traditions had also been robbed of their backbone, their best senior NCOs for training depots in their homeland.

The traditional British and Canadian view has been that two insufficiently trained, inexperienced, non-honogeneous, poorly equipped, and badly led Canadian militia battalions, with morale problems and hastily filled with new recruits, were sent to join highly effective battalions of the regular Imperial and Indian armies. Apparently this was not the case; with the exception of one battalion, the existing garrison and its Canadian reinforcements all suffered from similar problems. This alone would call for a reexamination of the existing history of the fall of Hong Kong. The exception was the Middlesex Regiment. They too had suffered losses in manpower and leadership and very few of them were, by December 1941, born within the "Sound of Bow Bells," their traditional Cockney recruiting ground. But their role as a machnine gun battalion rather than an infantry battalion, the extraordinarily high caliber of the regular officers remaining, the fact that they were not involved in the disastrous mainland defense, and the "family" feeling of the "Die Hards," into which the replacements were quickly incorporated, all seem to have contributed to a different attitude. Their regimental history is as full in its account of their performance as that of others is brief, and understandably. It is interesting therefore that their most distinguished surviving Hong Kong soldier, Christopher Man, who rose to the rank of Major General in the postwar army, should ask that his opinion of the Canadian troops be conveyed for any future reconsideration of the defense of Hong Kong. In comparison with the existing garrison, he points out "that the Canadians appeared fit, confident... and unusually well equipped by our standards."[15] This unprejudiced observer seems to bear out the conclusion that the previously accepted comparison between the Canadians and the Imperial and Indian battalions they were to join is a false one.

Such is the historical record as it stands. Yet there is a consensus among those who normally are most helpful in historical research that this unfortunate affair should never be reopened. For there are some who were involved in some way in the negotiations leading up to the despatch of the expeditionary force, the defense itself, and the consequences of the fall, who still agree with the Canadian Chief of the General Staff (Foulkes) when he told his minister a decade after the battle, and nearly twenty years ago, that "unless this case is reopened these regrettable circumstances can remain in oblivion."[16] The reasons for these attitudes are often valid from the holder's viewpoint, being due to political and diplomatic rather than personal fears. For instance, the renegotiation of the Hong Kong treaty or its present defense posture conceivably could be complicated by a critical reexamination of the only serious attack upon the colony and the British failure to withstand it.

So long after the event, it is doubtful if historians will ever be able to discover the "whole" story, if that is possible about any event. Still, given the availability of those who are willing to cooperate, and in view of the limitations imposed upon the official historians by time, space, and subject matter, there is a very strong case for a reconsideration of the Hong Kong Affair above and beyond the official and regimental histories. This is particularly true for Canada. It holds implications for Canadian political and legal history, the history of modern China, and the history of the imprisonment and its aftermath in the postwar pressure group that grew up around that shared memory.

Yet a history of the Hong Kong expeditionary force and the campaign must not be designed to reveal all solely in order to sell books or enhance a historian's reputation. The history of no military campaign is complete without substantial consideration of its importance in international affairs and its political and economic background. The Hong Kong Affair brings into question Mackenzie King's wartime leadership and the measures his government took to ensure that, in the interests of national unity, no challenge, however constitutionally sanctified, should be allowed to weaken its hold while the war continued. When George Drew publicly threw down the gauntlet for the Opposition, his private and parliamentary challenges over what he regarded as the Hong Kong fiasco were brushed aside and he was held on charges of treasonable activities against the Canadian state in time of war. The legal aspects of the Affair are as important historically as the constitutional ones. A safeguard of Canadian liberties, the Supreme Court, in the person of the Chief Justice, was first cajoled and then bullied into a Royal Commission which was designed to whitewash on grounds of raison d'etat. The Royal Canadian Mounted Police were used by the government to gather evidence for that enquiry. They were apparently instructed to ensure that those who could give evidence contrary to the government's position (that all possible had been done to provide an adequately trained and equipped force for the task envisaged, garrison duty where there was little chance of attack) were to be not only ignored but prevented from testifying.[18] These examples should suffice to show that the very nature of Canadian government in the war years must be reexamined in the light of the Hong Kong Expedition and the Royal Commission which followed it.

The importance of the British failure at Hong Kong in modern Chinese history is also underestimated by Canadian and British historians, although not by Chinese historians. To give one easily accessible example, Chan Lan Kit-Ching, lecturer in history at the University of Hong Kong, writing on the "Hong Kong Question during the Pacific War 1941-1945" in the *Journal of Imperial and Commonwealth History*, points out what a traumatic shock the fall of the colony was to the Nationalist Chinese government's will to continue the war as a fighting ally against Japan. The reluctance of the Kuo Min Tang forces to actively oppose the Japanese is too well known to repeat here. Suffice it to say there was reason to believe that this dormant posture would change as a result of Pearl Harbor, which brought into the war the allies China needed to go over to the offensive. However, "China was quickly disillusioned" about the power and the will of her new allies,

> particularly Britain, whose deeply entrenched position in South East Asia collapsed rapidly in face of the Japanese invaders. Hong Kong in fact started the ball of Chinese disappointment rolling fast. China was willing to write off Britain's loss of her two capital ships, the *Repulse* and the *Prince of Wales*, three days after Pearl Harbour, as "no more than bad luck which would be speedily made good. . . " But before the year was over, China's

greatest shock was the fall of Hong Kong. The Chinese were both surprised and worried: surprised because what they had hitherto regarded as a strong British base should be occupied by Japan with such ease; and worried because not a few of them, particularly those in senior government positions, had been using the colony as a refuge for their wives and children. Resentment more often than not followed the trail of disappointment and fear.[19]

The fall of Hong Kong assisted the Nationalist Chinese along the road toward accepting military stalemate with Japan. The tendency toward the policy of hoarding what military resources were available for a massive strike at the internal enemy, the Communists, rather than the Japanese, was strengthened.

As a result of the failure to defend Hong Kong adequately, Britain found it (in Chan's words) "most difficult to defend her position in Hong Kong. Intense pressure and devastating criticism were directed at her by both China and the United States. . . . one of the common questions asked in the United States about the Pacific War was why should Americans die to protect British Imperialism in places like Hong Kong." Some Canadians and Britons may have been asking the same question. As Chan points out, as a result of her failure and the resulting criticism, "Britain seriously considered giving up the colony."[20] Furthermore, there is reason to believe that Mao's views of the West's will to defend its interests in China, as expressed in the late 1940s to the cautious Stalin, confronted by a U.S. military giant with nuclear capacity, were influenced by the circumstances surrounding the fall of Hong Kong. It's possible that the "paper tiger" was born on Christmas Day 1941. Finally, what one might call the Pearl Harbor-Singapore syndrome establishes the narrowness of current military historians' approach to the history of World War II, if one considers the fall of Hong Kong alongside those disasters. Their repercussions and scale have led military historians into the attitude that the other simultaneous Japanese attacks on objectives in the Far East, including that on Hong Kong, were and remain sideshows. Did the fall of Singapore have such internal and external consequences for China as did that of Hong Kong? Do their ghosts still haunt international relationships involving successor governments as the spectre of Hong Kong does?

In Canadian domestic history, too, there are important longer-range reasons for major reconsideration. For example, the efforts to strike a Canadian East-West and Anglo-French balance, even in a force as small as that at Hong Kong, have contemporary parallels. The survivors tend to project a unity they perhaps did not quite possess in fact. The French-Canadian who played the most influential role in the history of "C Force" was

General Vanier, then in command of Valcartier Camp, from which the Rifles were despatched to the West Coast for their final destination. His evidence may show that a reexamination of the Hong Kong expedition can now better estimate the Canadian state of preparedness and effectiveness in 1941. When asked by the Royal Commission if he had passed the troops fit for their duties in Hong Kong, General Vanier replied that he had followed the normal procedure for a General Commanding at that time, he had stood on the saluting base and taken their eyes right as they marched out of his care toward the battleground.[21]

There had been in Hong Kong something of the state of mind of World War I. The Royal Scots, as the mainland defense battalion, had been told, before the 1940 debacle in France and the low countries, that they should not only prepare defensive lines based on those around the Ypres salient of over twenty years before, but that they should practice trench raids. This kept them busy under the instruction of a few very old soldiers who had experience of this kind of warfare, until Dunkirk made the practice of this tactic look as ridiculous as it was.[22] Perhaps Canada in 1941 had still not passed that stage in its thinking, particularly at the higher training, planning and staff levels in Ottawa. This is one of the claims to prominence of the Hong Kong expedition in Canadian military history: in many of such ways it is evidence of such a conclusion.

There are other unusual aspects of the fall of Hong Kong that are not adequately covered by the existing literature. First, the whole question of imprisonment is outside the scope of the official histories, but not outside the scope of the military historian. The leadership potential in the Royal Rifles and the "family" feeling was similar to that in the Middlesex regiment; the Winnipeg Grenadiers were equivalent to the Royal Scots. This in itself is interesting: seldom does one get the chance to analyse whole "units", particularly Canadians, of men going into captivity alongside a variety of Britons and Imperials, enabling us to examine the strengths and weaknesses of each. In the opinion of the doctors from the camps, the British regiments' long experience of the East, whatever it might have done to their fighting spirit, had brought them to terms with the normal standard of living, particularly in nutrition, in the East. This was unlike the Canadians, who were largely from rural or semi-rural areas. Despite the Depression, they were physiologically accustomed to a much higher standard of living than the British troops. It appears that there was a survival pecking order based upon past experience. The civilians at the top knew the level at which countless millions in China had existed for generations and understood that their level of physical well being, from which the will to live is directly derived, was no worse than that of those millions and generations. The British military fell into the middle category, and the Canadians, although they had moderately been deprived at home in the Depression years, came a long way last.

That bottom place in the survival league led those who did survive to feel that they had been treated as less than [Canadian] humans. Among them, the rate of disablement, permanent or temporary, complete or partial, was higher than among their fellow prisoners. Add to this the attitude of the King government on their return. It saw them as living reminders of past misdemeanors which might still come home to roost, with serious potential political and legal consequences. As a result there grew up in the postwar period a pressure group with a special history. From the Canadian Hong Kong veterans King had little to fear, had he but known it, yet from the beginning they were to be kept out of the public eye and not permitted to share in the heady fruits of victory. Colonel Price, the commanding officer of the survivors, en route homewards from the United States was warned not to give any comments to the press. Simultaneously, the award of the Pacific Star, which incidentally would have involved the payment of large sums of supplementary pay to add to the arrears of pay already due the survivors, was denied. The cold chill of the Canadian welcome was particularly galling after the warmth of the care offered by the American military authorities. The Minister of National Defense welcomed them home on the West Coast with words of pride on behalf of the Prime Minister. In the pregnant pause which followed, the perennial rear rank voice expressed the sentiments of all. "Bugger the Prime Minister, what about the Pacific Star?"[23] It has been that way ever since.

King had one last hurdle to negotiate before he could forget the Hong Kong expedition: those very official histories criticized here. In 1948 their pending publication raised the specter of disclosure. Unfortunately for the historian, however, the same Prime Minister was still in office and his opposition was still George Drew. Drew's chance of reversing the roles were small, faced as he was with a Liberal Party which had led the nation to a great victory. Drew took every opportunity to embarrass the Prime Minister, but the pending publication of the history of the Hong Kong expedition held a special attraction for him. If he could now reveal the truth about the failure in wartime leadership that the expedition represented, and the truth about the role of the Supreme Court in the enquiry which followed it, he could not only raise doubts about his opponents' contributions to victory, but perhaps even show they had delayed it. At the same time, he could revenge himself for the humiliations he had suffered at King's hands in 1942.

King had to be more thorough this time in his rebuttal of the storm of criticism which was bound to burst about his head from Drew, for although the undeclared Cold War was in progress he did not have the extraordinary powers which had devolved upon him as Prime Minister in the recent declared war. Nor could he use the Government's security agencies to suppress evidence surrounding an affair nearly a decade old, one in which the now dreaded Russians did not make an appearance. The evidence shows that he charged the Chief of the General Staff and his staff to reexamine all the evidence surrounding the despatch of the Expedition, including that suppressed in 1942, so that all embarrassing opposition questions might be met with satisfactory Parliamentary answers. The conclusions of the Chief of the General Staff as transmitted through his Minister were those of an extremely able and experienced field commander and staff officer. Extremely detailed, they may be summed up in his own opening advice, "I would strongly recommend that every effort should be made to avoid reopening this Hong Kong [Affair]." Among the regrettable circumstances he identifies is the fact that there was "nothing to show that the Department of National Defense [in 1941] had a staff which could work out the pros and cons of accepting the British proposal as a calculated risk of war" also "any such lack of energy" on the part of the Staff of the Department of National Defense in despatching the expedition to Hong Kong would "certainly with 1 Cdn. Corps have presented suitable grounds for a field general court martial."[24] Drew, however, had no access to these conclusions. King was able to follow the advice of the Chief of the General Staff. In the interests of national security and friendship with allies, this time the common enemy being the Russians and their spies in Ottawa, he was again able to turn Drew's by now rather antiquated missiles harmlessly away in Parliamentary answers. The Hong Kong expedition was again shelved.

There is today a new generation of military historians. After the official historians came the flood of participant's memoirs, in which all military historians drowned together. It is now time for a general reconsideration of the record. The term revisionism is unsuitable; it smacks of self-serving and ideological commitment. But something has to be done, for new factors are now emerging in the history of World War II which require thorough reexamination of the whole course of the war. The leading contender is the Enigma-Ultra factor, which demands a reevaluation of the military reputations of the Allied Commanders. How much they knew, and when, of the enemies' strength and intention may well promote some upwards on the ladder of historical reputation and demote others. The Hong Kong case is an example of failure of foresight. There are encouraging signs that most of the historical establishment recognizes this. With the coming of a new wave of military history, our understanding of World War II is likely to be very different from that now generally accepted.

Notes

1. Foulkes to Hon. Brooke Claxton, 9 Feb. 1948 Archives, Directorate of History, Dept. of National Defence Headquarters, Ottawa, 111.13 D. 66, hereinafter referred to as D.N.D.
2. *Ibid.*
3. Price to Nicholson, Deputy Director, Historical Section (G.S.) D.N.D., Ottawa, 27 Jan. 1948, D.N.D. 352.019 (D1) 1. Emphasis added.
4. *Ibid.*
5. *Ibid.*
6. Extract from Maltby's official dispatch on the Hong Kong Operations, attached to Nicholson's letter to Price, 13 Jan. 1948, D.N.D. 352.019(D.1.)
7. Memorandum to Colonel J.F. Meiklejohn, C.I.E., Historical Section, Cabinet Office, "Concerning the 2nd battalion, the Royal Scots..." from Augustus Muir, Historian, the Royal Scots, dated 26th March 1955, as supplied to the author by Augustus Muir (Pseud).
8. J. Leasor, *Green Beach* (London, 1975), 249.
9. There may some hesitation as to these characterisations of the defence and fall of Hong Kong, but after close scrutiny with John Swettenham at the National War Museum, Ottawa, it was agreed that on balance they were correct. The Canadian contribution to the invasion of Kiska is well encapsulated in the first line of the chapter dealing with the episode in Brian Garfield's interesting account of the Aleutian campaign, *The Thousand Mile War* (New York, 1970), 327. There have been occasions, particularly in the First World War, when the Canadian expeditionary forces were not exactly victorious, but they hardly fall into the category of major defeats. The defence of such "outposts of Empire" as Bermuda was essentially passive; not without influence in the overall picture, but certainly with little interest for anything other than the diplomatic or "Grand Strategy" historian.
10. J.A. Ford, *The Brave White Flag* (London, 1961), and *Season of Escape* (London, 1963).
11. Tim Carew, *The Fall of Hong Kong* (London, 1960), and *Hostages to Fortune* (London, 1971).
12. Oliver Lindsay, *The Lasting Honour: The Fall of Hong-Kong, 1941* (London, 1973).
13. B. Prasad, ed., *Official History of the Indian Armed Forces in World War II* (Delhi, 1960).
14. Carew, *Fall of Hong Kong* (London: Pan Edition, 1976), 22.
15. Cited from "Some rambling jottings on my experiences during the Battle of Hong Kong" at the request of the author. General Man's use of the term "[over] confident" in the original should be taken in the context of hindsight. No infantry battalion can be so characterized before the fact of battle.
16. Foulkes, *op. cit.*, 1.
17. Oral evidence given to the author by the former clerk to the Supreme Court.
18. Evidenced by letters from Drew's legal counsel in Vancouver, Brigadier Clarke, now in the Clarke Collection at the British Columbia Provincial Archives, Victoria, B.C.
19. Chan Lan Kit Ching, "The Hong Kong Question during the Pacific War (1941-5)", *The Journal of Imperial and Commonwealth History*, II, (1973), 57.
20. *Ibid.*, 73-74.
21. R.L. Kellock's copy of the *full* proceedings of Royal Commission... To inquire into and report upon the organization, authorization and dispatch of the Canadian Expeditionary

Force to the crown colony of Hong Kong," now in the DND Archives. Evidence given by Vanier, 715-719.
22. Oral evidence of Royal Scots Regimental Secretary (Col. B.A. Fargus) given to the author, Edinburgh, 1975.
23. Oral evidence given by Brigadier J.H. Price to the author, Montreal, 1974.
24. Foulkes' report and its supporting evidence, *op. cit.*

PERCEPTIONS OF MILITARY HISTORY

Learning Military Lessons from Vietnam: Notes for a Future Historian

Joseph J. Ellis
Mount Holyoke College

The Vietnam War has not yet become history. It lurks in that middle distance between present and past, where journalists regard it as passe and most historians consider it too recent for historical eligibility. When speaking of historians, of course, one should be careful, since they seldom agree on anything, much less reach a consensus on the time when events become history. One of my colleagues, for example, claimed that nothing which he could remember personally could qualify for history. For him, I take it, history begins somewhere in the late 1930s, although for myself, according to this definition, it would begin during World War II. Edmund Morgan once said, only half jokingly, that everything before 1800 is history and everything after that date is current events. Most historians, it seems to me, would agree that a history of the recent American involvement in Southeast Asia is not yet possible. Not enough time has passed. Despite the *Pentagon Papers*, not enough documents have been declassified. Too many of the participants and the ideological controversies remain alive to allow for that most murky of conceptions, historical perspective.

The soldier-scholars at the Office of the Chief of Military History, it is true, are already generating the official military history of the war, but these fat volumes are likely to prove useful as source books for future historians rather than as histories per se. Publishers, it is also true, have issued volumes purporting to provide the contemporary history of the Vietnam years, but the term "contemporary history" is inherently contradictory. Whether the Vietnam experience has had a dramatic and enduring influence on United States foreign policy or on the morale of the American Army, whether the United States government will make more judicious decisions about the proper scope of the "national interest" because officials have learned what Townsend Hoopes called "the limits of intervention", such are questions to which we cannot yet have answers. Indeed, we cannot know if they are the right questions. Advocates of a more pessimistic and fatalistic historical tradition, the disciples of Oswald Spengler and Arnold Toynbee, would surely argue that Vietnam is a symptom rather than a lesson, that it marks the beginning of the end of the American Empire, an event that reflects the waning of American global influence, the moment when the United States began its gradual but inevitable slide into the trash heap of history. Our modern-day Spenglers would scoff at questions which implied that nations are free to learn and choose. But Spenglerian pessim is not very fashionable in the United States, so we have heard little about "decline or fall", or about cyclical theories of history, except in undergraduate history courses during the beginning of the last week of the semester, when beleaguered instructors attempt to terrify their students into preparing for the final exam. For most historians and most Americans, the Vietnam memories are too recent and too raw to be examined either for lessons or cosmic warnings.[1]

Social scientists rush in where historians fear to tread. They have rushed in with specialized studies of civilian "decision-making" in the 1960s, psychological studies of Vietnam war veterans, computerized analyses of casualty rates, sociological studies of the membership of the war resistance movement.[2] Scholars who regard themselves as social scientists enjoy one splendid advantage over historians: they possess tools of analysis, such as mathematical models and data-producing devices, that enable them to comment confidently on various problems. The historians' major tool is time, a commodity not yet sufficiently abundant with reference to the Vietnam war. But even when the passage of time is sufficient, historians seldom find it possible to assert themselves with the confidence and authority that social scientists and military commanders customarily display. Historians tend to be acutely sensitive about the limits of their own intervention into controversial areas of inquiry; they tend to back away from hot issues until they have cooled down. This sense of intellectual limitation, in certain forms a laudable quality that promotes a seasoned and enduring brand of knowledge, can also breed excessive humility and caution; such scholarly caution partly explains why historians did not play a major role in the formulation of American policy in Vietnam.

I am reminded of a story that a West Point officer who served as staff member on the National Security Council during the 1960s once told.

> There was a scholarly type who was supposedly an expert on the history and culture of Southeast Asia and who was part of the policy-making task force in 1968. Whenever he was asked for his recommendations, all this historian had was complexities, nuances,

more problems. The rest of us didn't know dogshit about Vietnamese history, but when the Bundys or the Kissingers asked for recommendations, we had solutions. We may not have analyzed everything as deeply, but we had solutions. And the guys who needed the solutions said, "Hey, this historian's got no solutions." Frequently our solutions didn't work, but it always took a while to find out and by then the NSC guys had new problems that they wanted answers to.³

Lacking solutions or answers, convinced that America's Vietnam experience is not yet history, feeling a bit like the abovementioned "scholarly type" who could only see complexities and nuances, this historian would nevertheless like to risk a few tentative generalizations, mostly on the military policies pursued in Southeast Asia. These are offered, in Carl Becker's phrase, "without fear and without research," as the distilled musings on contemporary affairs by someone trained as a soldier and as a historian. The distant past is usually considered sufficiently unimportant to be left to historians, but contemporary events are too important to be left to social scientists, just as war is too important to be left to generals. Even though Vietnam is a very recent war, it is time for historians to begin the process of transforming it into history.

The Vietnam war has spawned a formidable literature designed to reveal what are called "the lessons of Vietnam." Since historians are supposedly the guardians of the past's wisdom, this seems to be the safest place to begin.

At a press conference in May, 1975, President Ford was asked what, in his judgment, were the lessons the United States had learned from its ill-fated involvement in Southeast Asia. "We have learned several lessons," the President responded. "The lessons of the past in Vietnam have already been learned—learned by Presidents, learned by Congress, learned by the American people—and now I think we should have our focus on the future." Not a single reporter asked the President what "the lessons of Vietnam" were. It was as if they understood that the ritualistic references to hard-earned "lessons" were not to be taken seriously, or at least not literally.⁴

How, then, should they be taken? It is difficult to know. Secretary of Defense James Schlesinger issued a public warning to the north Koreans in 1975 in which he asserted that "the United States government has learned the lessons of Vietnam" and that "unprovoked hostility against American troops in South Korea would be met with military force," "What did he [Schlesinger] mean?" asked Mary McCarthy. "That the next time we will use nukes? It all goes to show that the bruised leopard cannot change its spots." During the same year Senators Frank Church and Dick Clark invoked "the lessons of Vietnam" to oppose United States involvement in the Angolan civil war.[5]

In June of the same year the *New York Review of Books* asked some of its regular contributors to provide readers with incisive versions of their opinions on "the lessons of Vietnam." George Kennan announced that "the lessons of Vietnam are few and plain: not to be hypnotized by the word 'communism' and not to mess into other people's civil wars where there is no substantial American interest at stake." Robert Lowell, however, concluded that the lessons were too complex and ambiguous to grasp at this time and that Americans should not worry about learning lessons so much as kneel down and give thanks that the nightmare is over. Stuart Hampshire wrote that "Americans must learn to distrust chiefs of staff and military leaders as being liable always to be wrong about American policy." William Appleman Williams disagreed: "There is no way—repeat, no way—to saddle the military with the responsibility. . . True enough, the military was ready to serve as a true believer, but it did not define the world in such ways and it did not then cast policy in military terms." Williams thought the real culprits were the civilians in the State Department and the Pentagon. Sheldon Wolin was less interested in finding villains than in discovering ominous and cosmic ironies: "What all this means is a dramatic and qualitative break in American history. We have moved from a society of free choice to a society shaped by necessity. . . by some terrible irony we have been forced to enter history a second time, in this our bicentennial year. The first time we entered proclaiming our independence and liberty, the second time concealing our dependence and servitude." John K. Fairbank surmised that "the root cause of our Vietnam failure was . . . the profound American cultural ignorance of Asian history," but he was less than sanguine about what lessons could be drawn from the failure. "The lessons of history are never simple," he mused. "Whoever thinks he sees one should probably keep on with his reading."[6]

Fairbank's caution was the exception, although it was reminiscent of Albert Wohlstetter's warning issued in 1968, at one of the first Vietnam post-mortems. "Of all the disasters of Vietnam," said Wohlstetter, "the worst may be the 'lessons' we'll draw from it."[7] During the early 1970s, the United States was awash in pseudo-history, learning the lessons of Vietnam, the lessons of Watergate, sprinkling this depressing mixture with yeasty observations on the more uplifting meanings of the American Revolution. One could not pick up a newspaper or magazine without confronting Santayana's dictum: "Those who do not learn from the past are condemned to repeat it." For over a year, the best-selling book in the United States was David Halberstam's *The Best and the Brightest*, a massive compilation of anecdotes in which readers were shown how the parade of civilian and military leaders did not learn from their past mistakes and marched unknowingly into the Vietnam quagmire.[8] Halberstam's monstrous text satisfied the craving for quick wisdom; it played to the populist belief that American problems were traceable to an elitist strain of leaders bred at Harvard and West Point, nurtured in the RAND Corporation and the State Department, and planted in Saigon. Most of all, it fed on the lesson-learning mood of the time, the prevailing notion that history arranges itself into easily decipherable lessons, as if the gods piloted old biplanes that traced epigrams and axioms across the horizon. Future historians take note: Americans remained wedded to the wistful idea that there was an immediate "use" to which the Vietnam experience might be put. In this sense the innocence and naive hubris that in part were responsible for America's Vietnam policies in the first place survived the war experience intact; history remained something subordinate to America's laudable intentions.

This was not true of the American soldiers who served in Vietnam. From 1972 to 1974, while doing research for a book on West Point, I conducted interviews with scores of career Army officers who had served two and sometimes three tours of duty in Southeast Asia.[9] Few of these officers were willing to talk about the lessons of Vietnam, or indeed to generalize at all on the basis of their experiences. They wanted you to know that there were, in fact, several Vietnams; that a great deal depended on when one was there, whether one served in the delta or highlands region, whether one was in the Americal Division or the 101st Airborne, whether an officer was a Colonel or higher, in which case he usually viewed the war from the third tier of helicopters, or was a company grade officer, in which case he lived and fought alongside the troops. By and large these were frustrated and fatalistic professional soldiers who exhibited an almost Tolstoyan sense of war. They believed that all efforts to make war intelligible to the uninitiated were hopeless, that the notion of power or command in battle was an illusion, that nobody comprehended what was going on, and that they were in the grip of vast historical forces beyond their or anybody else's control. Their observations were as diversified as the scholarly and journalistic post-mortems we have already surveyed. But there were common themes, several persistent refrains running through their ruminations that indicated shared obsessions and mutual grievances. The comments of these officers constitute what historians call "primary sources"; perhaps it is best to let them speak for themselves rather than risk forcing them into an interpretive Procrustean bed.

First, the impact of Vietnam on the self-image and the public image of the military in America:

> Since the 1960s, since our involvement in Vietnam, soldiering has not been looked upon as an honorable profession . . . There's a tendency to apologize for being a soldier. Now war is a terrible thing. It's a dreadful thing and always has been. People who prosecute wars are, and have always had to be, at least part of the time, pretty terrible people. I don't think it's war that has changed much, or soldiers, but public knowledge and awareness. What we're doing is all on television. Americans liked to think of war in terms of two knights on a horse jousting with a lance. Some guy got knocked off and you picked him up. Now they know it's B 52 raids and napalm and artillery. I don't think Americans thought of soldiers in this context before. I find myself apologizing for my role in the war, not because I think this war is wrong, but because now everybody knows it's a terrible thing.[10]

This lament makes it easier to understand the blatant hostility between the military and the press during the war, when reporters, especially television reporters, exposed the unchivalric face of war to millions of viewers each night. "All members of the press are sons-of-bitches," said the normally restrained General Creighton Abrams to a member of his staff. "And if your father is a member of the press, he is a son-of-a-bitch too."[11]

Next, the belief that the military was betrayed in Vietnam, an American version of the German army's "stab in the back" theory of World War I:

> The Army is being set up to take the rap for the failures in Vietnam which were the direct and inevitable result of civilian decisions and indecisions. In Vietnam, it was seldom the prerogative of the military man to define his own mission. It was always imposed on him. The whole slant of the Vietnam experience has been vertical, from top to bottom, and we've always been on the bottom. The Army was never given a specific and accomplishable mission in Vietnam because McNamara and Johnson and their people never resolved the problem of whether they wanted to risk war with China. And down there at battalion and company level the commander shakes his head, pats his guys on the ass and says 'Go get 'em.' But they don't know what they're supposed to get. Then, when the great American public gets fed up, it's the guys with uniforms on who have blood on their hands.[12]

This point of view is widespread among professional soldiers. It is haunting;

some may even find it eloquent. It helps explain why Americans who apparently agree that Vietnam was a catastrophic national blunder frequently do not really agree at all. For one group regards the mistake to have been military involvement and the other group thinks the mistake was the limitation of the military involvement. More importantly, there are several crucial and faulty assumptions buried in this quotation, assumptions which we will need to scrutinize if we are to understand the significance of the military experience in Vietnam historically.

Last, a comment on the antagonism between senior and junior grade officers, or, perhaps, between soldiers who see themselves, to use Morris Janowitz's terms, as "managers" and those who see themselves as "heroes":

> There's a difference between a body count and a list of your own casualties, and there's a difference between all the goddamn numbers and holding a man in your arms when he dies. The guys who "orchestrated" and "managed" from the air conditioned trailers in the rear, they were the real enemy, not the VC. Too many trailer park Colonels never had the experience of war except as an abstraction... The guy in the command post who's not ridden on a chopper with recent casualties and hasn't seen the way the wash from the chopper causes the skin on the dead men to quiver like they were still alive, well, that guy, if he ever had that experience, wouldn't be able to "war-game the problem" any more.[13]

What are future historians to make of evidence such as this? Quite a bit, I suspect. War, so Sherman said, is hell. Few civilized persons would disagree with the thought, and Americans are not exceptional in this regard. From Stephen Crane's *The Red Badge of Courage* to Joseph Heller's *Catch-22*, American fiction has exposed the brutality and ignoble chaos of battle. After World War II, Samuel Stouffer's classic study, *The American Soldier*, rendered the glorified accounts of war and soldiering ridiculous by showing how unheroic, uncommitted, and terrified men at war actually were. Nevertheless, because the mass of Americans have always gone to war persuaded that they were waging a moral crusade against slavery, the Hun, Fascism, Communism or tyranny in some other form, war and soldiering have always retained a heroic or chivalric quality. And in the past 150 years only one conflict, the Civil War, has been fought on American soil, where the destructiveness and horror of the carnage could be experienced directly. In the 1960s, technological development, especially the development of television, confronted all classes of Americans with nightly pictures of burning bodies, severed limbs, and unspeakable atrocities. It was as if every evening, over one-half of the American population was required to read out loud from *All Quiet on the Western Front*.

It is possible that prolonged exposure to scenes from Vietnam only whetted the appetites of some Americans for more violence. But even if this is so, the violence of combat and war was no longer susceptible to glorification or ennoblement. Here it is essential to distinguish between anti-war and anti-military sentiment. While the televised reports from Vietnam did create what army officers called "an image problem" for soldiers, there is strong evidence that explicitly anti-military sentiment began to ebb even before the war ended. Applications to all the service academics rose steadily after 1972 and a nation-wide statistical study conducted in 1974 showed that the military received the highest rating when Americans were asked which of fifteen public service institutions were performing most effectively. But anti-war sentiment, as distinct from anti-military sentiment—the heightened public awareness of what military men are required to do—has endured. Its most visible manifestation has been the creation of an army composed entirely of professionals. This is an understandable but potentially disastrous response to the Vietnam trauma. The termination of the Selective Service Act and the establishment of an all volunteer army, direct consequences of the Vietnam experience, marked an end to a very old tradition: the ideal of the citizen-soldier, the belief that service in wartime was an obligation of citizenship. Although virtually all studies of the all volunteer army have indicated that it is likely to be less representative of and responsive to popular opinion, more expensive, more jealous of its own prerogatives, more xenophobic—in other words more likely to repeat some of the most grievous mistakes of Vietnam—it represented a way to shield the majority of Americans from exposure to war and warriors. Civilians, if they were to remain civilized, could not be trained in the ways of war.[14]

If Jacques Van Doorn is right, Vietnam was only the occasion, not the cause, for the establishment of a wholly professional army. Van Doorn is a Dutch sociologist whose major interest is military affairs and who, unlike most sociologists in the United States, does not seem anxious to divorce social science from the study of history. At the center of his vision is a perspective usually associated with social historians like Charles Tilly and Lawrence Stone. According to Van Doorn, the Vietnam war only accelerated a process already set in motion by larger, long-term forces. He argues that Vietnam may be remembered as the last war in which the United States used citizen-soldiers in battle, because the very conception of "a nation in arms" has become obsolete in the parliamentary democracies of the West. As Van Doorn sees it, "the Western world had now reached a historic stage in which the mass army as an institution is on the way out." One of the reasons for this change is technological: the introduction of tactical nuclear weapons and sophisticated electronic weapons systems requires highly trained specialists and smaller, more professional, and highly mobile troop detachments. More basically, Van Doorn points out that large armies of conscripts came into existence with the French Revolution and the national revolutions of the early nineteenth century. He asserts that it is no accident that this was also the era that witnessed the flowering of industrial capitalism, nationalism, democratic political institutions, and the belief in popular participation in and control of government. At this critical juncture, however, Van Doorn's analysis becomes clouded. His central point would seem to be that there is a direct and inevitable connection between mass armies filled with citizen-soldiers and a particular stage of social, economic, and political development. We have now moved beyond that stage, he suggests, into a post-capitalistic, post-democratic world and all the institutions and ideologies associated with that older world are fading away together.

Van Doorn's explanation works quite well for Western Europe, where Britain ended conscription in 1962 and both the Netherlands and France are currently following suit. Whether it works as well for the United States is less clear. One might be permitted to withhold assent as long as the Soviet Union and the People's Republic of China continue to defy the historical trend as Van Doorn sees it and maintain the largest mass armies in the world. But whatever its shortcomings, Van Doorn's interpretation is intriguing because it does something that desperately needs doing: it sets the analysis of military problems in a broad historical context that stretches across centuries and national borders, establishing a genuinely historical perspective on contemporary military affairs.[15]

Vietnam has given war a bad name among mainstream Americans and has helped to segregate the military from the rest of American society. It has also raised again a fundamental question which was originally raised in the 1950s: in the nuclear age, what is the purpose or mission of the United States army? This question has not been posed directly so far. We are still in that period of time following most less-than-successful wars when the participants defend their decisions, generals dictate their memoirs with an eye toward posterity's judgment, and investigative reporters search for culprits. The historian must listen for deeper tones amidst the cacophony of shrill accusations and special pleadings.

Recall the earlier plea of an officer who worried that the army was "being set up to take the rap" for failures which were the direct result of misguided civilian decisions. He complained that the army never had "an accomplishable mission" in Vietnam because civilian policymakers had not decided whether they wanted to risk war with China or the Soviet

Union. This assertion is not accurate. One point on which civilian officials of the Kennedy, Johnson, and Nixon administrations all agreed was that American military involvement in Southeast Asia should not be allowed to escalate into a war with one of the other superpowers. Military men are certainly correct to charge that civilians must bear a lion's share of the responsibility for what happened in Vietnam, especially because they ordered the insertion of American troops in the first place. But in the end, although there is plenty of blame to go around, culpability is not the main issue. The main issue here, or at least the main strategic issue, is the army's understanding of its mission and its inability to adapt that understanding to the special conditions posed by limited war, especially the kind of limited, unconventional war found in Vietnam.

Two incidents illustrate the implicit assumptions which controlled American strategic thinking. Many others could be cited, but these two occurred at a critical phase in the war and became precedents for later decisions. The first occurred in the spring of 1965. Two battalions of Marines landed at Danang on March 8, charged with the mission of defending the airfield there. At this early stage of the war there were only about 12,000 American troops in Vietnam and they were specifically limited to advisory roles, meaning that they were intended to provide tactical and technical advice to South Vietnamese troops. They were not allowed to fight except when attacked by Viet Cong forces; they were specifically forbidden to initiate offensive operations against the enemy. Less than a month after their arrival, however, the Marine battalions were authorized to conduct "search and destroy" missions within a fifty mile radius of Danang. These offensive operations, which produced heavy American casualties and led to a request for two additional Marine battalions, were justified as responses to Viet Cong attacks on Danang. Offensive operations, in other words, became the only way of carrying out the original defensive mission. In less than a month, American soldiers had been transformed from military advisors, stationed in Danang for security, to combat troops whose primary mission was to serve as an infantry strike force.[16]

The second incident occurred in the summer of the same year, 1965. In June, General Westmoreland requested the buildup of American ground forces to approximately 200,000 men. Despite the offensive operations of the Marine units around Danang, official American strategy had remained advisory and defensive up to this time and Defense Department memoranda continued to speak of American involvement as a "police action." Then two things happened. First, the Johnson administration not only granted Westmoreland's request, but also authorized him "to commit U.S. troops to combat, independent of or in conjuncion with GVN forces. . . when in

[Westmoreland's] judgement, their use is necessary...."[17]

If one were looking for the precise moment when civilians abdicated their responsibility for American policy in Southeast Asia, this would be the place to find it. For Westmoreland here received *carte blanche* to use American troops in Vietnam as he saw fit. To give an American military man so free a hand, it turns out, was tantamount to the adoption of an offensive strategy designed to destroy enemy forces and the logistical bases from which they obtained their supplies. This is precisely what Westmoreland did. He called for "the resumption of the offensive by U.S./F.W.M.A. forces... in high-priority areas to destroy enemy forces... and... the defeat and destruction of the remaining enemy forces and base areas" by the end of 1966.[18] In sum, during the spring and summer of 1965, the American armed forces on the ground in Vietnam adopted the strategy which Russell Weigley has called "the American way of war." That strategy called for offensive combat operations aimed at the physical destruction of the enemy and his means of support, the tacit acceptance of the principle of annihilation and the goal of total military victory. American military leaders imposed this goal on themselves, despite the fact that military victory was not the aim of American policy in Southeast Asia and, as ten years of fighting subsequently demonstrated, was never a realistic possibility. Why did this happen?[19]

The answer to this question requires a backward glance, first at the relatively recent past and then further back. During the late 1940s and early 1950s, it is now clear, the United States institutionalized its Cold War doctrines and committed itself to a foreign policy based on a belief in the need to contain communism, which was then regarded as a monolithic movement with headquarters in Moscow. This containment theory, usually associated with George Kennan and Dean Acheson, continued to shape the thinking of American policy-makers into the 1960s and led directly to the decisions that involved the United States in Southeast Asia. Few scholars, no matter what their ideological persuasion, would dispute the conclusion that Vietnam represented an application of foreign policy principles formulated in the aftermath of World War II.[20] But these were also the years when civilian and military thinkers initiated a heated debate over the military strategy most appropriate for a global foreign policy. In the early 1950s advocates of "massive retaliation," also called the "New Look," were in the ascendance; they argued for reliance on U.S. nuclear capability to deter Soviet aggression and they minimized the need for a build up of conventional weapons and armed forces. In this debate, leading spokesmen for the Army became passionate advocates for the alternative to a strategy of "massive retaliation," which denied the ground forces a major role.

Generals Matthew Ridgway, James Gavin, and Maxwell Taylor all wrote books urging the adoption of a strategy of "flexible response," in which the United States would compliment its nuclear arsenal with greatly strengthened non-nuclear forces specifically designed to oppose communist aggression in situations that did not justify nuclear confrontation. Bernard Brodie, Robert Osgood, and Henry Kissinger became the most prominent civilian advocates of "flexible response," but the Army immediately embraced the same arguments. Limited wars, i.e., non-nuclear wars fought for limited and carefully defined objectives, became the Army's chief *raison d'etre*.

By the early 1960s, as the situation in Southeast Asia deteriorated in the wake of the French defeat and withdrawal, the American army had come to view Vietnam as an arena in which it could demonstrate its continued relevance in the nuclear age and its effectiveness as America's major military instrument in limited war. Vietnam was an opportunity to show that a strong American army was not an anachronism. It was a testing ground, not just for new weapons and new tactics, but for the usefulness of the Army itself. General Westmoreland's eagerness and optimism during the critical months early in 1965, then, were not the consequences of a diabolical character but the predictable products of recent history.[21]

If we step back even further into the past, another feature of the strategic terrain also comes into focus. Russell Weigley has written a perceptive and comprehensive history of strategic thought in America in which he has identified the controlling assumption of American military leaders from Sherman to Westmoreland: that the sole purpose of war is decisive victory and that this is achieved by, in the motto of the infantry, closing with and destroying the enemy. General Douglas MacArthur summed up this assumption neatly when he said "There is no substitute for victory... War's very object is victory." When Weigley calls this "the American way of war," he implies that it has roots in the unique natinal experience of the United States. It seems more correct, however, to describe it as a way of thinking about war that accompanied the rise of mass armies throughout the West in the late eighteenth and early nineteenth centuries. In Europe it appeared during the Napoleonic Wars. In the United States it became a full-blown strategic philosophy after Grant and Sherman assumed military control of the Union armies during the Civil War. It is not so much an American way of war as what social historians would call a modern way of conceiving of war, an offensive and aggressive strategy that came into existence alongside industrial captitalism, nationalism, and political liberalism.

Two features of this strategy are worthy of special attention. First, it is a creature of history, the product of a particular time and a particular set of social conditions. This is critical because most soldiers believe that the principles of war are absolute; any strategy that does not have total victory as its end and physical destruction of the enemy as the means to that end does not fit their conception of war. Second, there is an affinity between the personal qualities and values essential for successful practitioners of this strategy and the personal qualities essential for entrepreneurs who are successful in a marketplace free of government regulation. This should come as no surprise. As Alfred Vagts put it in his classic *A History of Militarism*: "We may say that each stage of social progress or regress has produced military institutions in conformity with its needs and ideas, its culture as well as its economics." The psychology underlying modern military strategy crystalized during the blossom days of liberal capitalism in the West and reflects the deepest convictions associated with the free enterprise philosophy. For our purposes, the most salient feature of this psychology is the profound belief that the pursuit of profit in the marketplace or victory on the battlefield, once initiated, must not be constrained by any outside force.

No matter how efficacious this mentality once may have been, it is clearly an anachronism in the contemporary world. MacArthur's dictum to the contrary notwithstanding, the object of war is no longer total victory, a fact that MacArthur ought to have noticed on the banks of the Yalu in 1950. Unfortunately, outdated convictions have the capacity to endure long after the social conditions that gave rise to them have become history. Although the American army became a staunch advocate of limited war during the 1950s, it retained the older belief that war was a total contest to the bitter or victorious end. It never internalized the new assumptions inherent in the strategy of limited war, especially its indecisiveness, the emphasis it placed on defensive operations, and the wholly new goal of political stability rather than traditional military victory. Behind the steady escalation of American military operations in Vietnam, one does not discover stupidity, or conspiracy, or willful intransigence. One discovers soldiers acting in accordance with anachronistic but still compelling notions of war, serving as living proof of Faulkner's claim that "The past is never dead, it is never even past."

It should be clear that the problem here is not susceptible to rapid solution by means of a change in the curriculum at West Point or at the Command and General Staff College at Fort Leavenworth. Like the dominant American attitudes toward blacks, or toward the supremacy of what is called the free enterprise system, American soldiers' attitudes towards war have been tempered in the furnace of time. The adoption of a new kind of military strategy, of new convictions about war that are appropriate to present and future conditions, will prove just as difficult and time-consuming as the adoption of new attitudes towards economic development and personal fulfillment appropriate for a post-capitalistic, limited growth society. Perhaps the most worrisome feature of the all-volunteer army is that it encourages soldiers to insulate themselves from civilian society and allows them to cling tenaciously to outmoded visions of the profession of arms. It certainly puts an increased burden of responsibility on civilian officials to impose restraints on military operations, restraints which the soldiers will surely perceive as unjustified.

But now I am looking toward the future rather than the past, behaving as a prophet rather than a historian. Instead of bequeathing a legacy of predictions to future generations of historians, it is more sensible to leave them with a few questions which might serve as directives for their investigations. First, what impact did the Vietnam war have on popular American attitudes towards war? Second, was the establishment of an all-volunteer army the right or wrong lesson to have learned from the Vietnam experience? Third, did the psychology, even the world-view, of American soldiers in Vietnam reflect a fundamental misunderstanding of the nature of limited warfare, a misunderstanding rooted in an older but forever lost world?

There remains a fourth implicit question: what role, if any, should the historian have in the formulation of American military policies? This must remain an open question, in part because any answer I provided would be self-serving, in part because historical wisdom, at its best, tends to assert the inevitability of certain courses of action and the limited range of options open to officials (who prefer to see themselves as decision-makers rather than prisoners of the past), and in part because the historian's equivocation, the "on-the-one-hand, but then on-the-other-hand" mode of reasoning, would probably drive sane and stable officials to the edge of lunacy. Nevertheless, one cannot read through the *Pentagon Papers* without wondering if the Defense Department or the National Security Council might have benefitted from the presence of an officially designated gadfly whose primary responsibility was to say "Yes, but...." Historians certainly possess the requisite contrariness and flair for the idiosyncratic insight to make them perfect choices for such a position. Although very few historians possess the verve required to make nuance look definitive, most historians are superb critics of someone else's allegedly foolproof proposals. Since the usual practice is to regard the past as a grab-bag of convenient examples designed to illustrate the wisdom of policies already decided on for reasons that have little to do with history, a person with even a cursory knowledge of the past possesses a remarkable capacity to wreak havoc. Imagine the consternation that might have seized Dean Rusk if his incessant and derogatory references to Munich and appeasement were challenged by someone who pointed out that Vietnam was in Asia, not Europe, and Ho Chi Minh was not Hitler. Or imagine the dead silence that might have followed Walt Whitman Rostow's strenuous argument that strategic bombing of Germany during World War II justified the bombing of Hanoi and Haiphong, if someone recalled that voluminous strategic studies indicated that the aerial bombing raids on Germany were inconclusive. Historians have the capability and the responsibility to make themselves useful nuisances.

Notes

1. Maurice Matloff, general ed., *American Military History* (Army Historical Series, Office of the Chief of Military History, Washington) represents official history at its best. The best of the "contemporary histories" are Roger Hilsman, *To Move a Nation: The Politics of Foreign Policy in the Administration of John F. Kennedy* (New York, 1968); Arthur Schlesinger, *A Thousand Days: John F. Kennedy in the White House* (Bostom, 1965); Chester Cooper, *The Lost Crusade: America in Vietnam* (New York, 1970). The crucial primary source is Neil Sheehan, et. al., *The Pentagon Papers as Published in the New York Times* (New York, 1971). On learning from military history, see John Shy, "The American Military Experience: History and Learning," *Journal of Interdisciplinary History* (Winter, 1971), I, 205-28.
2. Summaries of the extensive social science literature can be found in Adam Yarmolinsky, ed., *The American Military Establishment* (New York, 1971); Sam Sarkesian, ed., *The Military Industrial Complex: A Reassessment* (Beverly Hills, 1972); John Lovell, "No Tunes of Glory: America's Military in the Aftermath of Vietnam," *Indiana Law Journal* (Summer, 1974), XLIX, 698-717; "Symposium on the Vietnam Experience," *Armed Forces and Society* (Spring, 1976), II, 339-420.
3. Interview with West Point officer, January, 1973.
4. Transcript of the press conference, *New York Times*, May 7, 1975.
5. *Ibid.*, May 28, 1975; McCarthy quotation, *New York Review of Books*, June 12, 1975, 25.
6. *New York Review of Books*, June 12, 1975, 23-33.
7. Quoted in Lovell, "No Tunes of Glory," *op. cit.*, 710.
8. David Halberstam, *The Best and the Brightest*, (New York, 1972). For a devastating critique of Halberstam's book, see Mary McCarthy's review, *New York Review of Books*, January 25, 1973, 3-12.
9. Joseph Ellis and Robert Moore, *School for Soldiers: West Point and the Profession of Arms* (New York, 1974).
10. Interview with Lt. Colonel Thomas E. Blagg, March, 1973.
11. Interview with West Point officer who served on General Abrams' staff, May, 1973.
12. Interview with West Point officer, August, 1973. The distinction between "heroes" and "managers" was first made in Morris Janowitz's *The Professional Soldier: A Social and Political Portrait* (New York, 1960).

14. U.S. President's Commission, *The Report of the President's Commission on an All-Volunteer Armed Force* (Washington, 1970); Martin Binkin and J.D. Johnston, "All-Volunteer Armed Forces: Progress, Problems, and Prospects," Report prepared for the Committee on Armed Services, United States Senate (Washington, D.C., 1973); Jerald Bachman and John Blair, "Citizen Force or Career Force: Implications for Ideology in the All-Volunteer Army," *Armed Forces and Society* (Fall, 1975), II, 81-96.
15. Jacques Van Doorn, "The Decline of the Mass Army in the West: General Reflections," *Armed Forces and Society* (Winter, 1975), I, 147-58; Morris Janowitz, "Military Institutions and Citizenship," *Armed Forces and Society* (Winter, 1976), II, 185-204.
16. Matloff, *American Military History*, 623; Sheehan, *Pentagon Papers*, 382-409.
17. Sheehan, *Pentagon Papers*, 411-12.
18. *Ibid.*, 463.
19. Russell Weigley, *The American Way of War: A History of United States Military Strategy and Policy* (New York, 1973), 441-78, for a detailed discussion of military strategy in Vietnam.
20. Two good accounts of post-World War II diplomacy are Walter Lafeber, *American, Russia, and the Cold War 1945-1967* (New York, 1967), and Stephen Ambrose, *Rise to Globalism* (Baltimore, 1971). For a critique of "revisionist" history, see Robert Tucker, *The Radical Left and American Foreign Policy* (Baltimore, 1971).
21. Weigley, *American Way of War*, 399-440; Bernard Brodie, *Strategy in the Missile Age* (Princeton, 1959); Robert Osgood, *Limited War: The Challenge to American Strategy* (Chicago, 1957); Henry Kissinger, *Nuclear Weapons and Foreign Policy* (New York, 1957). Books by army officers on this subject include Matthew Ridgway, *Soldier: The Memoirs of Matthew B. Ridgway* (New York, 1957); James Gavin, *War and Peace in the Space Age* (New York, 1958); Maxwell Taylor, *The Uncertain Trumpet* (New York, 1960).

U
42
.M43
1982